AF332923

DNA-BASED NANOSCALE INTEGRATION

Proceedings in this Series of International Symposiums

Year	Symposium	Publisher	ISBN
2006	DNA-Based Nanoscale Integration	AIP Conf. Proceedings Vol. 859	978-0-7354-0357-4
2004	DNA-Based Molecular Electronics	AIP Conf. Proceedings Vol. 725	0-7354-0206-X
2002	DNA-Based Molecular Construction	AIP Conf. Proceedings Vol. 640	0-7354-0095-4

To learn more about these titles, or the AIP Conference Proceedings Series, please visit the webpage **http://proceedings.aip.org**

DNA-BASED NANOSCALE INTEGRATION

International Symposium on
DNA-Based Nanoscale Integration

Jena, Germany   18 – 20 May 2006

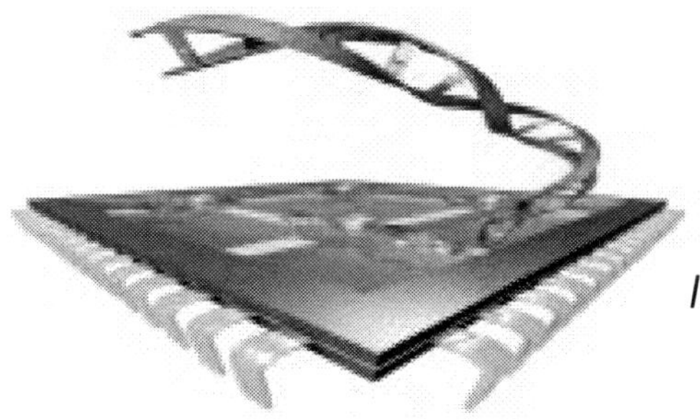

EDITOR
Wolfgang Fritzsche
*Institute for Physical High Technology,
Jena, Germany*

SPONSORING ORGANIZATIONS
Main Sponsor: Volkswagen Foundation
Institute for Physical High Technology - IPHT
Fonds of the Chemical Industry - FCI
Foundation for Technology, Innovation and Research in Thuringia - STIFT

Melville, New York, 2006
AIP CONFERENCE PROCEEDINGS ■ VOLUME 859

Editor:

Wolfgang Fritzsche
Institute for Physical High Technology (IPHT)
P. O. Box 100 239
D-07702 Jena
Germany

E-mail: fritzsche@ipht-jena.de

L.C. Catalog Card No. 2006933425
ISBN 978-0-7354-0357-4
ISSN 0094-243X

Printed in the United States of America

CONTENTS

MICROSCOPIC CHARACTERIZATION

APPENDICES

PREFACE

DNA Nanotechnology coming of age

The utilization of DNA as a tool in molecular manipulation on solid substrates was the focus of the conference "DNA-Based Nanoscale Integration" May 18-20, 2006 at the IPHT in Jena (Germany). Although envisioned already decades ago, the last years showed a tremendous progress that is evident in the broad application of these approaches not only by a few specialized groups, but on a wider scale. So DNA nanotechnology is changing from its niche existence as a technique only for specialists to a broader application, especially in nanotechnology. The appearance of DNA nanotechnology as a standard technique beside a variety of other biomolecular and physical approaches promises further progress in the manipulation at the nanometer and single molecule scale, and is especially interesting for the integration and combination of these novel conjugates with rather traditional technical environments (such as microsystems and microelectrodes) in order to fully explore the potential for future applications.

Wolfgang Fritzsche

IMMOBILIZATION

Fabrication of DNA Mediated Devices: Alignment of single DNA and the 1-D pattern of DNA self-assembly

Yongli Mi*, Kris Wan, Bryan Wei, I-Ming Hsing, Chi-Ming Chan

Department of Chemical Engineering, the Hong Kong University of Science and Technology, Clear Water Bay, Kowloon, Hong Kong

Abstract. In this study, we investigated the DNA anchoring technique on gold substrates and the DNA stretching method by molecular combing. The stretched DNA molecules were visualized by the fluorescent microscope. The stretched DNA was also evidenced by the silver reduction along DNA backbones with hydroquinone in both alkaline and acidic conditions. The scanning electron microscope (SEM) shows that the unstretched DNA molecules result in random silver coils, whereas the stretched DNA molecules give the straight silver wires of more than 10 μm long. The molecular combing technique was successfully demonstrated in bridging single DNA molecule over the gold microelectrodes, which is observed by the fluorescent microscope. The DNA anchoring on the gold surface was enhanced by the aminoethanethiol modification and by oligonucleotide hybridization. The aminoethanethiol modification of the electrode surface was analyzed by a Physical Electronics PHI 7200 ToF-SIMS Spectrometer and a reflectron ToF analyzer. A more sophisticated DNA 1-D structure was also achieved by the DNA self-assembly and visualized by transmission electron microscope (TEM) with the negative staining. These achievements suggest a useful approach of converting biological macromolecules to potential applications in micro electronic industry.

Keywords: DNA, molecular combing, metallization

PACS: 87.14.Gg, 81.16.Dn, 81.07.Nb, 81.16.Rf, 85.35.–p, 87.15.Mi

INTRODUCTION

There has been a tremendous interest in recent years to develop concepts and approaches for miniaturized self-assembled systems for microelectronics and optical applications. Research has been extended to seeking inspiration from biological sciences, using living organisms, with functional elements that are of molecular dimensions, as the components of ultra-small electronic circuits and computers. Many recent studies have been directed to apply the biological macromolecules to develop new techniques in nano-circuitry.

Since the discovery of double helical structure of DNA by Watson and Crick in 1953, DNA also flourished in the materials field rather than being restricted in the genetic and biochemical studies. One of the reasons for using DNA in the material

CP 859, *DNA-Based Nanoscale Integration: International Symposium,* edited by W. Fritzsche
© 2006 American Institute of Physics 978-0-7354-0357-4/06/$23.00

world is the self-assembly ability by molecular recognition, which can create highly organized and predictable structures (scaffolds) to meet miniaturization for micro electronics. By depositing metal particles on DNA scaffolds, micro- and/or nano-scale wires can be formed for applications in the microelectronic industry.

In early 1990's, Seeman and coworkers laid a foundation for using DNA templates in pattern constructions.[1] Among the applications of DNA in nanoelectronic devices, the self-assembly of DNA conjugated nanoparticles have received the most attention in recent literature.[2] Mirkin et al. and Alivisatos et al. were among the first to describe self-assembly of gold nanoclusters into periodic structures using DNA. Significant work has been reported in the last 10 years towards the DNA-mediated assembly of artificial nanostructures. The concepts have recently been extended to metallic nanowires and nanorods.[2] In 1998, Braun and coworkers successfully grew DNA bridges between two electrodes and deposited silver onto the DNA to obtain a nanoscale conducting wires. They also described a method of assembling highly luminescent filaments of poly(p-phenylene vinylene) (PPV) using DNA as templates for possible optical applications.[3] Apart from using DNA as a template for nanostructure fabrication, scientists have also gained advantages from using self-assembled proteinaceous structures of cytoskeleton to fabricate tubular nanostructures. Mertig et al. in 1998 reported parallel fabrication of identical nanotubes by electroless metal film deposition onto microtubules derived from self-assembled protein filaments.[4] Mertig et al. in the same year also described another method of fabricating highly oriented nanocluster arrays using the bacterial surface layer (S layer) of Sulfolobus acidocaldarius as the protein template.[5]

All above examples can demonstrate that biological macromolecules may provide a solution to electronic manufacturing at micro-scale.[6-11] In this work, we want to extend the potential of DNA molecules to the applications in microelectronic components.

EXPERIMENTAL

Gold Electrodes

Gold electrodes of 5 μm apart, as demonstrated in Figure 1, were fabricated on glass slide and silicon wafer by standard photolithography. The gold electrodes were first cleaned with H_2SO_4/H_2O_2 solution (7:3 volume ratios) in an ultrasonic bath at room temperature for 10 minutes, followed by an ultrasonic treatment with nanopure water for another 5 minutes. Immediately after that, all the gold electrodes were rinsed thoroughly with nanopure water.

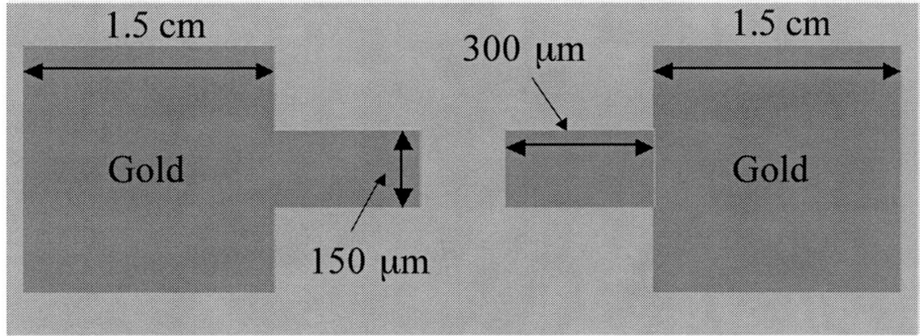

Figure 1 The topographical view of the microelectrode. The gap between the electrodes is 5 µm.

Stretching of λ-DNA Molecules by Molecular Combing

Stretched λ-DNA Observed by Fluorescence Microscope

The DNA and YOYO-1 complex solution was prepared according to the following procedures. 20pM λ-DNA in Tris-acetate EDTA (TAE) buffer (40 mM Tris-base, 20 mM acetic acid and 1 mM EDTA, pH 8) was stained with YOYO-1 at a 5:1 ratio (base pairs per dye molecule). The DNA solution was gently added to the dye solution and equilibrated with YOYO-1 for at least 12 hours in order to avoid any precipitation and to ensure homogeneous dye binding.

A drop of prestained λ-DNA solution (2 µL) was deposited on a clean glass cover slip treated with aminopropyltriethoxysilane (APTES).[12-15] Another clean cover slip with a smaller diameter was placed on the top of the droplet to force it to spread out. They were kept in a vacuum oven over night. All the samples were kept in amber containers throughout the whole drying process. After the λ-DNA solution in between the two glass surfaces dried, samples were stored in a desiccator for 2 hours prior to further treatments. The dehydration step before fluorescence imaging was necessary to prevent DNA from rapidly detaching from the glass surface. With this dehydration step, stretched and fixed DNA molecules remained on the surface long enough to allow the acquisition of fluorescence imaging.[16] The silanized glass surface with the stretched DNA molecules was flushed with nanopure water for 30 seconds to wash away all the unbound DNA molecules. To verify the alignment of DNA molecules attached to the glass substrate after molecular combing, fluorescence microscope was used. An inverted microscope (Axioskop 7082, Carl Zeiss), equipped with 100x oil-immersed objective, 12V halogen lamp with filters, image intensifier (Photometrics Sensys, CTS Chinetek Scientific), was used for fluorescence imaging of the DNA molecules. An imaging measurement software (SPOT), version 3.5, Diagnostic Instruments, Inc, Michigan, USA, was used to perform measurements of the fluorescent micrographs. A stage micrometer (Leica, Germany) was used to calibrate the 100x oil-immersed objective on the inverted microscope prior to measurements.

Custom oligonucleotides were purchased from Invitrogen Corp. (www.invitrogen.com). DNA strands were purified by electrophoresis; bands were tailored out of 15% denaturing polyacrylamide gels and eluted in a solution containing 500 mM ammonium acetate, 10 mM magnesium acetate, and 1 mM EDTA. The 9 oligonucleotides were mixed (equal mole) in 20 mM Tris (pH 7.6), 2 mM EDTA, and 12.5 mM $MgCl_2$ (TAE/Mg). The final concentration of each DNA strand was 1.0 µM and the final volume was 50 µL. Oligonucleotide mixture was cooled slowly from 95 °C to 4 °C in a heating block for approximately 56 hours to facilitate hybridization.

Images of transmission electron microscope (TEM) were obtained on a Jeol JEM 2010 TEM. Direct observation of DNA structures was carried out by pipetting 10 µl of annealed DNA solution onto a copper grid with porous carbon film. Copper grid was placed on a piece of filter paper, so that the buffer solution could be removed by the filter paper. The sample grid was then moved to a silicon wafer. Negative staining was adopted by first pipetting 10 µl of 2% uranyl acetate solution onto the sample grid for 5 min. The excess liquid was then removed from the sample by lightly touching an edge of filter paper to the edge of the grid. The sample grid was put in the desiccator over night before the TEM examination.

Bridging of DNA over Electrodes by Aminoethanethiol Modification

Preparation of Self-Assembled Monolayer (SAM) of Aminoethanethiol on Gold Electrode

A clean gold electrode was soaked in a 5 mM aminoethanethiol hydrochloride/ethanol solution for 10 hours at ambient temperature.[17-18] The surface of the electrode was washed with ethanol and nanopure water, and then dried by pure nitrogen gas. After these treatments, self-assembled monolayer of aminoethanethiol with primary amino groups was formed on the gold electrode surface. The aminoethanethiol modification of the electrode surface was analyzed by a Physical Electronics PHI 7200 ToF-SIMS Spectrometer (ICI, Wilton, UK) equipped with two ion guns (Cs+ for high mass range surface spectroscopy analysis and 69Ga+ for spatially resolved imaging) and a reflectron ToF analyzer. The image acquisition was carried out in both positive and negative modes using a 25.0 keV gallium liquid-metal ion source, with the beam being scanned over an area of 200 x 200 µm (256 x 256 pixels) on the sample surface. In order to obtain the high spatial resolution and the easy control of surface charge stabilization, an ion pulse of about 50 nanoseconds (ns) was imposed and the beam size was about 0.5 µm. The signals were detected using a 256 stop time-to-digital converter (TDC) with 10 ns resolution. The complete image was acquired from up to a maximum of 200 frame scans using one primary ion pulse per pixel. In both modes of analysis, a low energy (0-70 eV) electron flood gun was used to achieve charge compensation and static conditions by keeping the total ion dose below 4 x 10^{12} ions/cm^2. The data were acquired in the so-called "raw data stream" mode in which the full spectral information at every pixel was stored to create a complete raw date stream for each image. An image from a selected peak (or a group of peaks) and the spectra from a selected image area can be created. All data acquired in this work were analyzed using the manufacturer's Physical Electronics ToF-PAK software.

Anchoring and Stretching of λ-DNA Molecules on the SAM-Coated Gold Electrodes

The treated gold electrode was covered by a 2 μL solution of 20 pM λ-DNA prepared in TAE buffer containing 0.1 M EDC prestained with YOYO-1[3]. The electrode was incubated in a homemade humid chamber at ambient temperature for 30 minutes. It allows anchoring of λ-DNA molecules to the gold electrodes by the formation of a phosphoramidate bond between the 5' phosphate group of λ-DNA and the amino group of the SAM coated gold surface. After incubation, a flow of the solution perpendicular to the electrodes was induced by micropipette suction of the solution.[3] The flow stretched the λ-DNA molecules from one gold electrode to the other electrode, bridging across the gap between the two electrodes. After drying, the electrode with λ-DNA was immersed in washing buffer (0.4 M NaOH and 0.25% SDS heated to 40 °C) for 5 minutes, then washed 3 times with the buffer solution and flushed with nanopure water to remove unbound λ-DNA molecules. The gold electrode was then ready for fluorescent imaging. Note that all the samples were covered at all times by aluminum foil or kept in amber containers to protect the samples from photo bleaching of the fluorophores. The result of DNA anchoring, between the two electrodes, according to this method, is illustrated in Figure 4.

Bridging of DNA over Electrodes by Oligo Hybridization

Immobilization of Oligonucleotides on Gold Electrode Surfaces

A 5.2 nL droplet of an aqueous solution containing 10 μM oligonucleotide A and 100 mM NaCl/25 mM sodium phosphate (pH 7.0) was added to the surface of one of the gold electrodes by using a microinjector (Stemi SV6 stereomicroscope, Carl Zeiss). Together with the micropipette, the microinjector was calibrated to repeatedly deliver uniform inoculating droplet diameter as small as 100 μm. Similarly, the other gold electrode was wetted with a solution containing 10 μM oligonucleotide B in 100 mM NaCl/25 mM sodium phosphate (pH 7.0) following the same procedures. The electrodes with oligonucleotides A and B, respectively were then incubated in a homemade humid chamber for 3 hours at room temperature, and then flushed with nanopure water to remove the excess oligonucleotides. The base sequences of oligo A and oligo B with a disulphide group at the 3' end are shown in Figure 5.

After rinsing the gold electrode surfaces immobilized with both the oligonucleotides A and B at the two electrodes, the electrodes were covered by a 2 μL solution of YOYO-1 stained λ-DNA (50 pM) prepared in 2 x SSC (300 mM NaCl/30 mM Na citrate, pH 7.0). The λ-DNA has two cohesive ends, one of which is complementary to oligonucleotides A, and the other to B.

Since the free thiol-modified DNA is liable to oxidative dimerization to form disulfide, resulting in a reduced probe immobilization efficiency. In order to eliminate this problem, both the oligonucleotide A and B were treated with dithiothreitol (DTT) prior to their immobilization. 0.1 volume of 1 M DTT was added to the probe solution and incubated at ambient temperature for 30 minutes. The DTT was then removed by a standard ethanol precipitation procedure. The solution was mixed thoroughly with 0.1 volume of 3 M sodium acetate (pH 5.6) and 3 volumes of ice-cold absolute ethanol, kept in a –80 °C freezer for 30 minutes, and centrifuged at 14,000 rpm for 15 minutes at 4 °C. The DNA pellet, thus obtained, was washed with 1 mL 95% ethanol. After centrifugation, the pellet was dried and re-suspended in water. Finally, the concentration of the oligonucleotide was determined and adjusted to 10 μM in 100 mM NaCl/25 mM Na phosphate (pH 7.0) as the oligonucleotide immobilization solution.

A drop of λ-DNA solution (2 μL) was deposited on a gold electrode surface across the gap. A clean cover slip was then floated on top to force the drop to spread. After the evaporation of the λ-DNA solution, the gold surface was flushed with nanopure water. In order to verify the alignment of DNA molecules attached to gold electrodes across the gap after molecular combing, fluorescence microscopy was applied. The result of DNA anchoring, between the two electrodes, according to this method, is illustrated in Figure 7.

Silver Ion Reduction on the DNA Backbones

Silver ion deposition and reduction were performed on DNA molecules on glass slides. The DNA molecules were first impregnated with silver ions using a 0.1 M $AgNO_3$ basic aqueous solution (ammonium hydroxide, pH = 10.5) at ambient conditions for 2 hours. Prior to the silver ion reduction, the silver ion/DNA complex was washed thoroughly by nanopure water in order to remove the unbound silver ions. The silver ion/DNA complex was then reduced using a basic hydroquinone solution (0.05 M, 10% sodium sulphite, ammonium hydroxide, pH = 10.5) to form small metallic silver aggregates bound to the DNA skeleton. The further development of silver reduction was accomplished by using an acidic solution of hydroquinone (0.05 M, citrate buffer, pH = 3.5) and silver ions (0.1 M) under dark condition.[3] The silver wires thus formed on the glass slides were submerged in 1 mM sodium thiosulphate solution for 10 minutes prior to thorough washing by nanopure water to stop the reaction. A control experiment of silver reduction, without using DNA as a template, was also performed. In order to visualize the morphology of the silver wires dwelled along the DNA backbones, observation was performed with a high-resolution scanning electron microscope (SEM). Samples were kept in a desiccator before use and were sputter-coated with a ~150 Å of gold film. The electron-acceleration voltage was 30 kV with a magnification range of 500–50,000.

RESULTS AND DISCUSSION

Molecular Combing of DNA

In order to find DNA some applications as the templates, DNA molecules should be firstly stretched. By a molecular combing technique discussed above, Figure 2 shows a fluorescent micrograph of λ-DNA molecules combed on a glass cover slip that was pre-silanized with APTES at pH 8.0 (40 mM Tris-base, 20 mM acetic acid and 1 mM EDTA, 20pM λ-DNA).

Figure 2 A fluorescent micrograph of λ-DNA molecules combed on glass cover slip that was silanized with APTES at pH 8.0 (40 mM Tris-base, 20 mM acetic acid and 1 mM EDTA, 20pM λ-DNA).

Construction of DNA Bridges between the Aminoethanethiol Modified Gold Electrodes

After investigating the stretching technique of long-chain DNA molecules on glass substrates, we tried DNA bridging over the aminoethanethiol-treated gold electrodes. Time-of-flight SIMS imaging was used to characterize the deposition of the self-assembled monolayer (SAM) of aminoethanethiol on the gold electrode surface. Images were generated after acquiring mass spectra. Figure 3 shows the images of the control gold electrodes (Figure 3A) and the aminoethanethiol-modified gold electrodes (Figures 3B-3D). The images were generated by combining the signals obtained from a number of the nitrogen containing ions generated from the fragmentation of the aminoethanethiol backbone. After the incubation of the electrodes in aminoethanethiol hydrochloride/ethanol solution and subsequently rinsing with ethanol and nanopure water, a homogeneously distributed layer of amino group appeared on the gold electrodes, c.f. Figures 3B-3D.

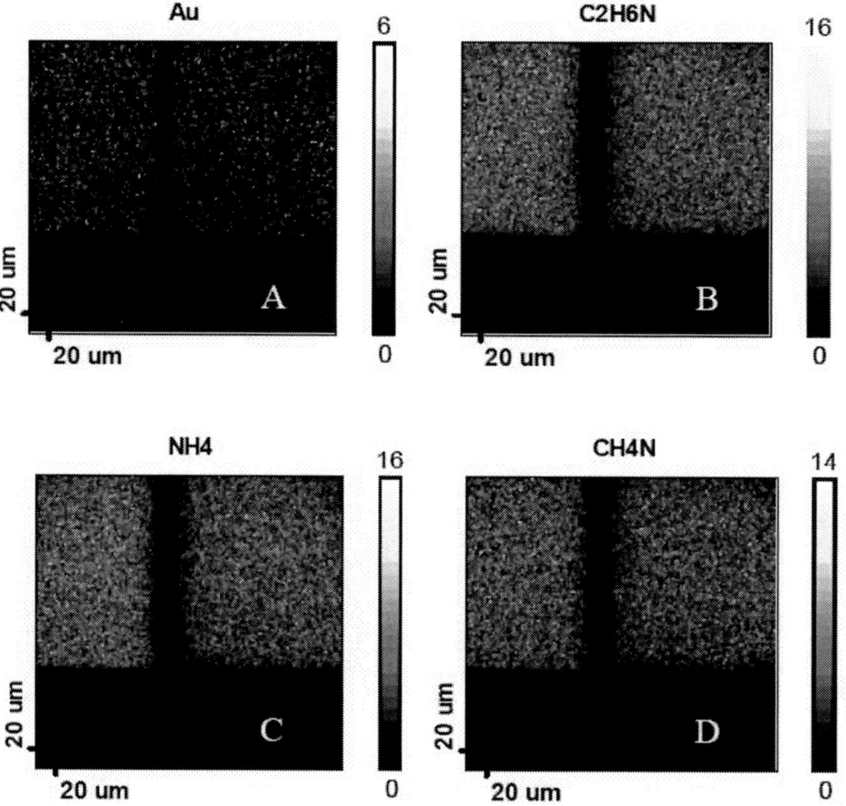

Figure 3 Time-of-flight SIMS images of the control gold electrodes (A) and the aminoethanethiol-modified gold electrodes which show strong N density on the electrodes (B-D).

Figure 4 shows a well stretched λ-DNA has been successfully bridged over the aminoethanethiol modified gold electrodes. The concentration of the λ-DNA was 100 pM (TAE buffer, pH 8.0). It shows that both extremities of the λ-DNA strand are found to anchor to the aminoethanethiol modified gold surfaces. The simple and convenient pipette suction method thus used ensured the stretching of the λ-DNA molecules in the flow direction that is perpendicular to the gold electrodes, which promises the reproducible construction of DNA bridges between gold electrodes. The amount of DNA bridges constructed between the electrodes can be changed by adjusting the concentration of DNA in the deposition solution; however, making electronic devices with exactly one DNA molecule per pair of the electrodes, reproducibly, is still a stochastic process. More studies are expected in developing the controlled number of DNA bridges over the electrodes. In this study, we observed one to several DNA bridges for the experiments described above.

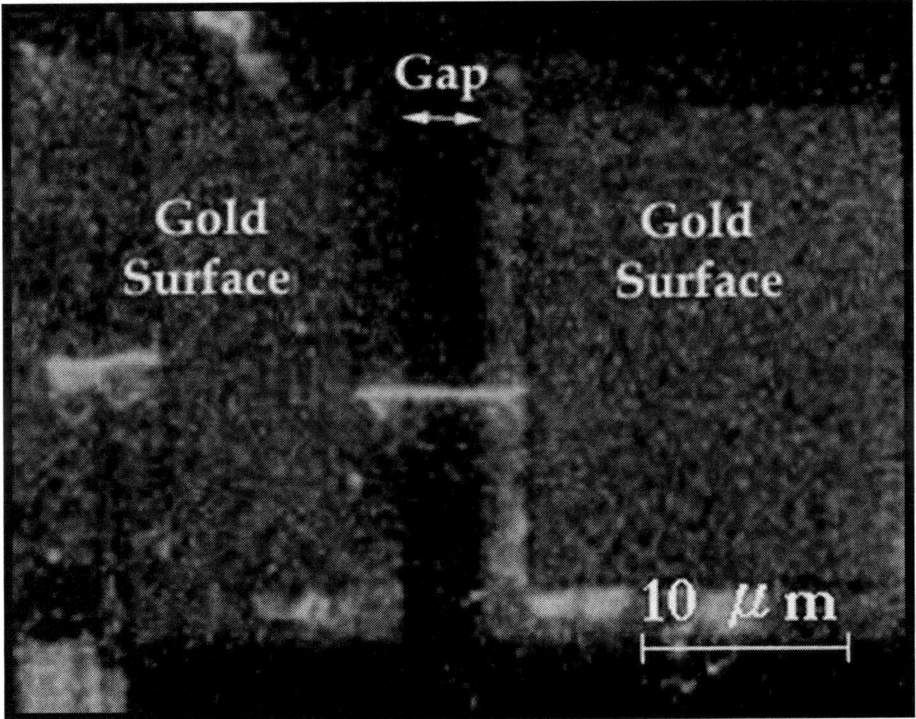

Figure 4 A fluorescent image of stretched λ-DNA molecule bind to the aminoethanethiol modified gold electrodes (20 pM YOYO-1 stained λ-DNA in TAE buffer, pH 8.0). The DNA bridge is formed between the two gold electrodes fabricated on a silicon wafer.

Construction of DNA Bridges between the Gold Electrodes by Oligo Hybridization

The method of using the chemisorption of thiol-modified DNA oligonucleotides onto a gold surface has proved to be a good technique for a stable and reproducible DNA-oligo immobilization[3,19]. We also applied this approach to construct the DNA bridges over the electrodes. The bridging scheme of λ-DNA over the two gold electrodes, by hybridizing its two ends with oligo A and oligo B, respectively, is schematically demonstrated in Figure 5. Figure 6 shows the optical fluorescence characterization of a droplet of fluorescently labeled oligo A immobilized on one gold electrode surface by microinjection, which supports the oligo attachment to the gold surface. Figure 7 shows the results of DNA bridging over gold electrodes. Figure 7a is the picture of the electrodes before DNA bridging. After DNA bridging, one finds, in Figures 7b and 7d, single DNA bridge over the two electrodes, and in Figure 7c, multiple DNA bridges. As described in the above section, the control of single DNA bridge over a given pair of electrodes is still problematic, which needs more future studies.

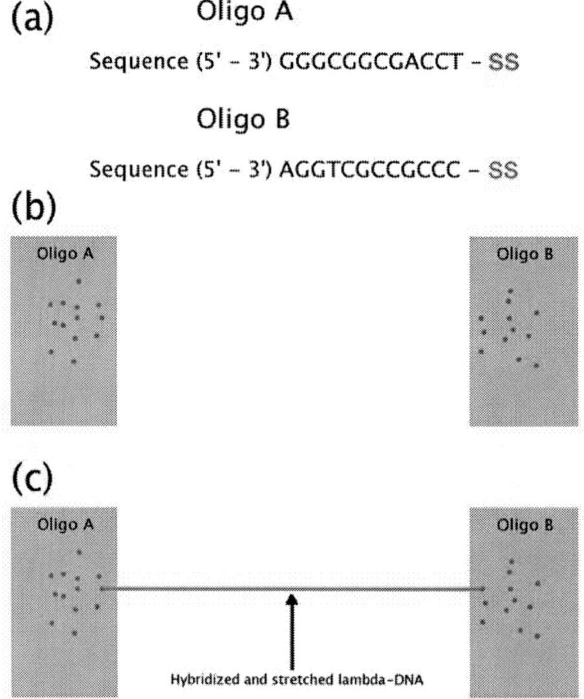

Figure 5 (a) The base sequences of oligo A and oligo B with a disulphide group at each of their 3' ends; (b) immobilization of oligo A and oligo B on gold electrodes by microinjection; (c) The bridging scheme of λ-DNA over the two gold electrodes by hybridization its two ends with oligo A and oligo B, respectively.

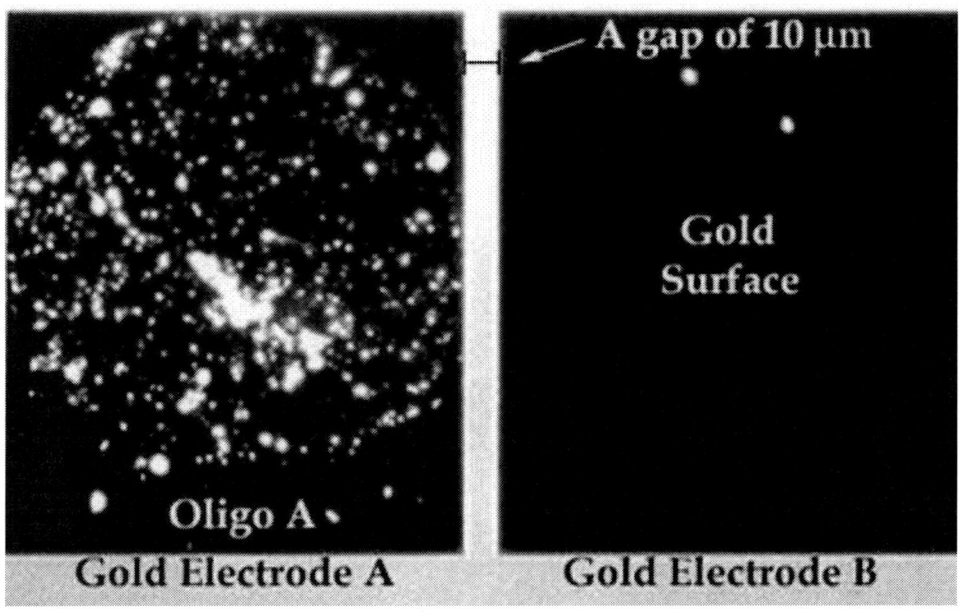

Figure 6   The optical fluorescence characterization of a droplet of fluorescently labeled oligo A immobilized on one gold electrode surface by microinjection, which supports the oligo attachment to the gold surface.

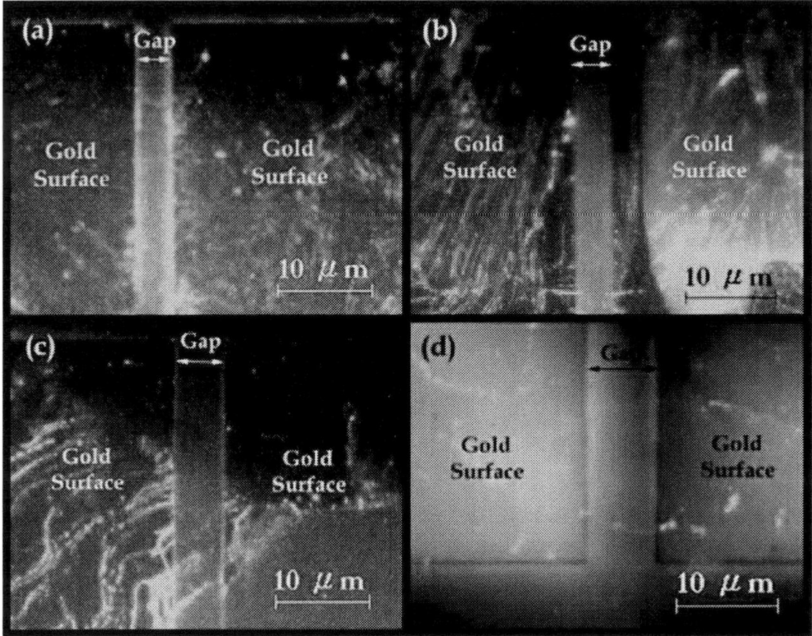

Figure 7   (a) the picture of the electrodes before DNA bridging. (b) and (d) single DNA bridge over the two electrodes. (c) multiple DNA bridges over the two electrodes.

Silver Reduction along the DNA Backbone

Electroless plating was applied for the metallization of DNA molecules stretched on glass substrates. This was accomplished by impregnating DNA with silver ions. Depending on the duration of the reduction process, well-separated silver clusters formed a continuous silver wire with a grain-like structure along the DNA backbones. Figure 8 shows that, without the DNA molecules, silver particles grow in randomly dispersed fashion on the substrates (Figure 8a). For the unstretched DNA molecules, silver reduction magnifies the random coiled nature of the long chain molecules, as shown in Figure 8b. Figure 8c also confirms the molecular combing is successful, which shows that the metallic silver particles can grow along the DNA backbone forming a silver wire.

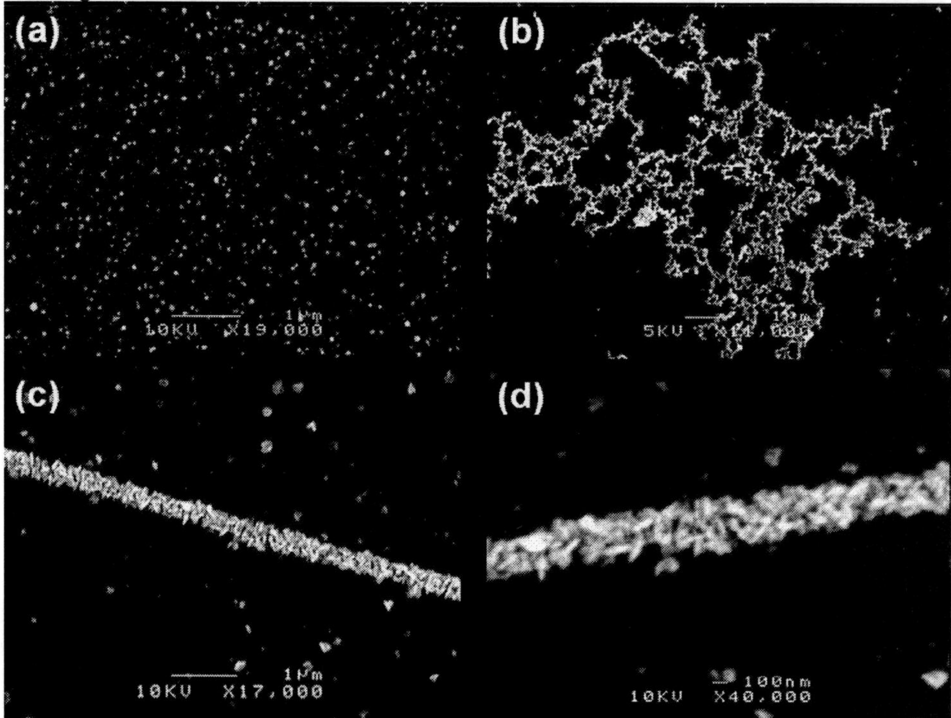

Figure 8 SEM micrographs of the silver reduction, (a) without DNA templates, silver particles randomly appear on the substrate, (b) with unstretched DNA molecules as the template, the silver particles grow in the pattern that resembles the random coils of the unstreched DNA molecules, (c) and (d) with stretched DNA molecules as templates at the magnification of 17,000 and 40,000, respectively.

Study of the Influence of Silver Ion Reduction Time

The silver ions form strong complexes with DNA during the reduction. Excess silver ions were removed by thorough rinsing. By controlling the reduction time, silver deposition occurs only along the DNA skeleton, leaving the substrate clean of silver. The reaction of DNA with silver ions was monitored by quenching the fluorescence signal in accordance with the observations by Braun *et al* and Richter *et al*.[3,20] The impregnation of DNA molecules with silver ions completes in 20 minutes depending on the amount of DNA molecules on the sample. Within a few seconds of the alkaline developing agent being added, nano-scale clusters of silver grew on the DNA skeleton. The silver deposition process was terminated when a trace of the silver metal was observed under the optical microscope after several times of washing. SEM observation shows that the silver coating on stretched DNA, after the early stage of reduction with alkaline solution, is only discontinuous as illustrated in Figure 9a. However, these silver aggregates on the DNA template act as catalysts and significantly accelerate the subsequent acidic reduction process to further deposit silver grains around the DNA scaffold. When the duration of the reduction process is extended, silver particles coalesce and finally develop a continuous metallic wire along the DNA backbone. Figures 9b and 9c show the silver wire formed with reduction reaction at different reduction time.

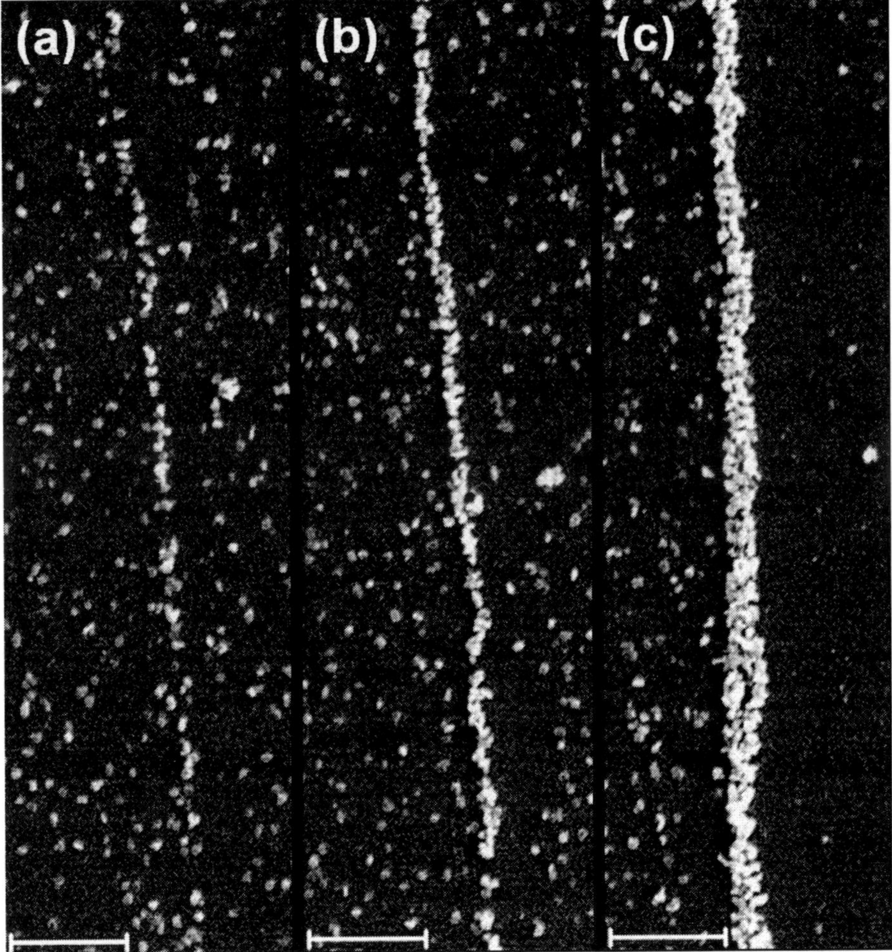

Figure 9 The progressive reduction of silver on the templates of stretched DNA molecules, (a) the alkaline reduction process, (b) the acidic reduction process for 5 minutes, and (c) the acidic reduction process for 10 minutes. (scale bar 1 μm).

1-D DNA self-assembly

New initiatives were also taken to develop versatile DNA nano wires. We adopted the 4x4 DNA tile reported by Yan et al[21] and kept two ends (top and bottom) inert and two ends (left and right) active as shown in Figure 10a, in which one can see the unpaired base groups ready for further assembly with each other to form a linear structure as shown in Figure 10b. Nine oligonucleotides were involved in building the 4x4 tile unit and the sequences of the nine oligonucleotides are shown in Figure 10a. The 4x4 tile units (Figure 10a) were firstly formed, and then, the crosses were linked one and another by desired sticky end (the hydrogen bonding) matching strategy to form a linear structure, as shown schematically in Figure 10b. The TEM image of the 1-D DNA self-assembled structure, thus obtained, is shown in Figure 11. The scale bar is 0.2 μm.

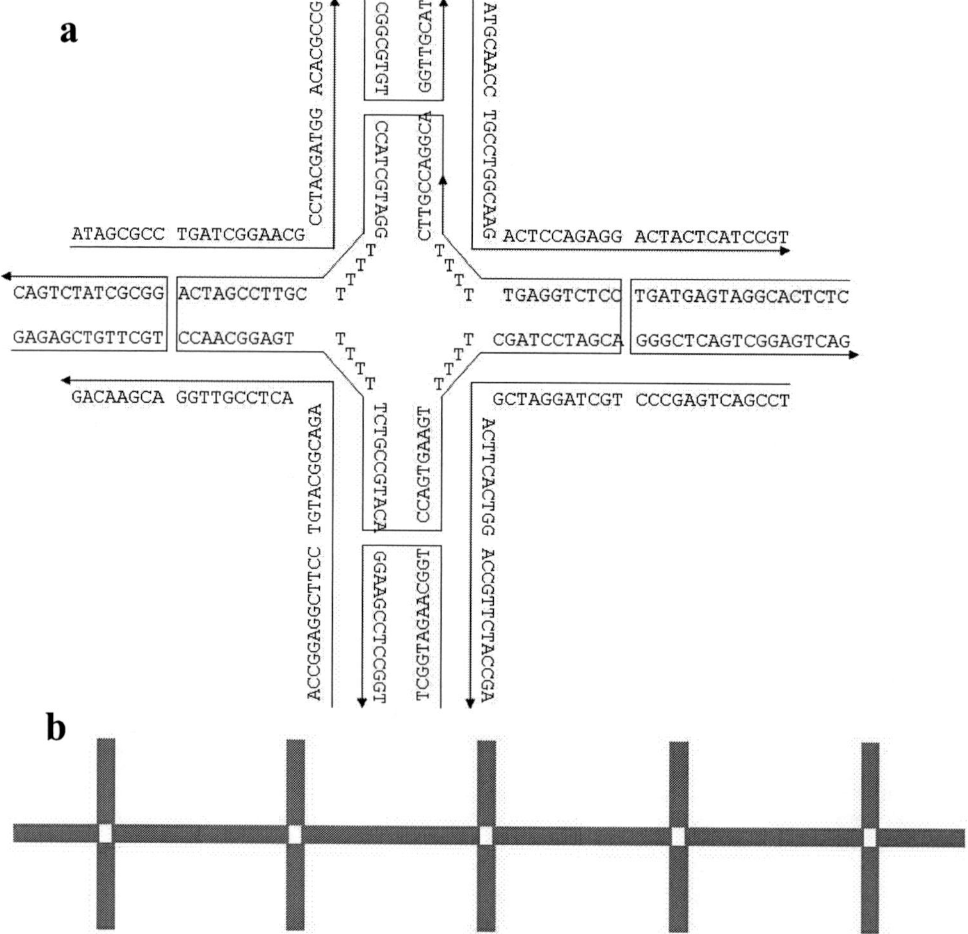

Figure 10   (a) The DNA 4x4 tile and the sequences of the 9 oligonucleotides. (b) The schematically diagram of the 1-D self-assembled structure by the DNA 4x4 tile.

16

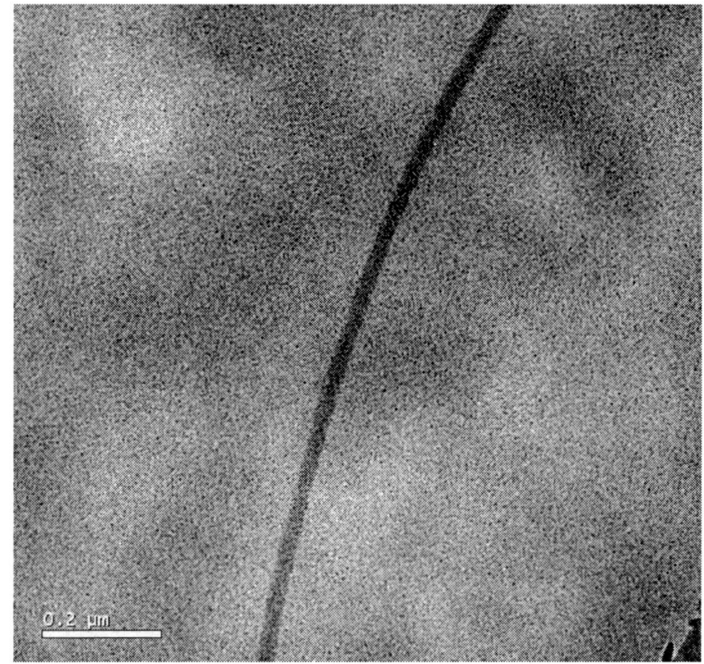

Figure 11   The TEM image of the 1-D DNA self-assembled structure.

CONCLUSION

In this work, DNA molecules are successfully engineered in the stretched state and self-assembled state. The long-chain double-stranded DNA molecules appear to be an attractive biomolecular material used as templates for nano-structure fabrication. The formation of silver wires involved in the micro-manipulation of DNA molecules in linear pattern, served as a template, followed by silver ion deposition. The fabrication of silver wires was then accomplished by silver ion reduction treatment which enabled the metallic silver aggregated along the DNA template. Deposition of silver ions into the DNA backbones followed by chemical reduction using hydroquinone in both alkaline and acidic conditions leads to the production of high quality silver wires with silver grains aggregated along the DNA scaffolds. Electron micrographs show that the silver wires formed with DNA templates are of more than 12 μm long and approximately 300-500 nm in width. These achievements suggested a useful approach in which natural found biomolecules can be found applications in microelectronic industry.

ACKNOWLEDGEMENT

The financial support of the RGC research grant, 602603, is greatly acknowledged.

REFERENCES

1. Seeman, N. C. *Nanotechnology* **1991**, *2*, 149-159.
2. Mirkin, C. A.; Letsinger, R. L.; Mucic, R. C.; Storhoff, J. J. *Nature* **1996,** *382*, 607-609.
3. Braun, E.; Eichen, Y.; Sivan, U.; Ben-Yoseph , G. *Nature* **1998**, *391*, 775-778.
4. Mertig, M.; Kirsch, R.; Pompe, W. *Applied Physics A* **1998**, *66*, S723-S727.
5. Mertig, M.; Kirsch, R.; Pompe W.; Engelhardt, H. *The European Physical Journal D* **1999**, *9*, 45-48.
6. Bashir, R. *Superlattices and Microstructures* **2001**, *29*, 1-16.
7. Sarikaya, M. *Proceedings of the National Academy of Sciences of the United States of America* **1999**, *96*, 14183-14185.
8. Winfree, E.; Liu, F.; Wenzler, L. A.; Seeman, N. C. *Nature* **1998**, *394*, 539-544.
9. Merkle, R. C. *Trends in Biotechnology* **1999**, *17*, 271-274.
10. Drexler, K. E. *Annual Review of Biophysics and Biomolecular Structure* **1994**, *23*, 337-405.
11. Service, R. F. *Science* **1999**, *286*, 2442-2444.
12. Krebs, M. R. H.; Wilkins, D. K.; Chung, E. W.; Pitkeathly, M. C.; Chamberlain, A. K.; Zurdo, J.; Robinson, C. V.; Dobson, C. M. *Journal of Molecular Biology* **2000**, *300*, 541-549.
13. Bensimon, A.; Simon, A.; Chiffaudel, A.; Croquette, V.; Heslot, F.; Bensimon, D. *Science,* **1994***, 265*, 2096-2098.
14. Yokota, H.; Johnson, F.; Lu, H.; Robinson, R. M.; Belu, A. M.; Garrison, M. D.; Ratner, B. D.; Trask, B. J.; Miller, D. L. *Nucleic Acids Research* **1997**, *25*, 1064-1070.
15. Allemand, J. F.; Bensimon, D.; Jullien, L.; Bensimon, A.; Croquette, V. *Biophysical Journal* **1997**, *73*, 2064-2070.
16. Bensimon, D.; Simon, A. J.; Croquette, V.; Bensimon, A. *Physical Review Letters* **1995***, 74*, 4754-4757.
17. Wan, W.; Lin, J.; Schwartz, D. C. *Biophyiscal Journal* **1998**, *75*, 513-520.
18. He, P. G.; Ye, J. N.; Fang, Y. Z.; Suzuki, I.; Osa, T. *Electroanalysis* **1997***, 91*, 68.
19. Johansson, A. and Stafstrom, S., Chemical Physics Letters, **2000**, 322, 301-306
20. Richter, J.; Seidel, R.; Kirsch, R.; Mertig, M.; Pompe, W.; Plaschke, J.; Schackert, H. K. *Advanced Materials* **2000**, *12*, 507-510.
21. Yan, H.; Park, H. S.; Finkelstein, G.; Reif, J. H.; LaBean, T. H.; Science 301, 1882 (2003).

Defined DNA immobilization for a DNA-based micro-nano integration

A. Csaki, A. Wolff, T. Schüler, R. Möller, G. Festag, R. Kretschmer, G. Maubach[a], W. Fritzsche

Institute for Physical High Technology, Albert-Einstein-Straße 9, 07745 Jena, Germany
[a] present address: IBN Singapore, 31 Biopolis Way, The Nanos #04-01 Singapore 138669

Abstract. DNA represents a highly promising molecule for the construction of a wide variety of materials and devices utilizing molecular units. Integration into today's microsystems technology provides access to these interesting but usually solution-based (and thereby difficult to control) constructs. A variety of techniques to realize this combination have been developed. Of special interest are techniques that are open for massive parallelization in order to provide sufficient numbers of test structures, but also to allow for a potential scale-up on the way to future applications. We describe and review here a selection of promising technical developments in this direction, addressing the integration problem.

Keywords: DNA nanotechnology, microstructures,
PACS: 81.16.–c, 87.14.Gg, 81.07.–b

INTRODUCTION

DNA is recognized as probably the most versatile molecule for molecular construction due to its unique properties including the complementarity-based binding capabilities and the large body of knowledge about it acquired in molecular biology. Numerous materials and devices based on DNA were proposed and some of them were even already demonstrated, consisting of DNA alone or in combination with additional materials such as proteins or nanoparticles. Although these structures are often prepared in solution, the standard microscopic characterization (such as AFM or sometimes fluorescence microscopy) occurs on solid substrates due to technical reasons. But solid substrates represent also a connection to the technical world, a world that reaches already the nanoscale (as evident in the IC production). Especially in the case of electrical characterization, but also for many other studies as well as future applications, a highly defined connection between both worlds – the macroscopic technical and the molecular nanoscale - is required.

CP 859, *DNA-Based Nanoscale Integration: International Symposium*, edited by W. Fritzsche
© 2006 American Institute of Physics 978-0-7354-0357-4/06/$23.00

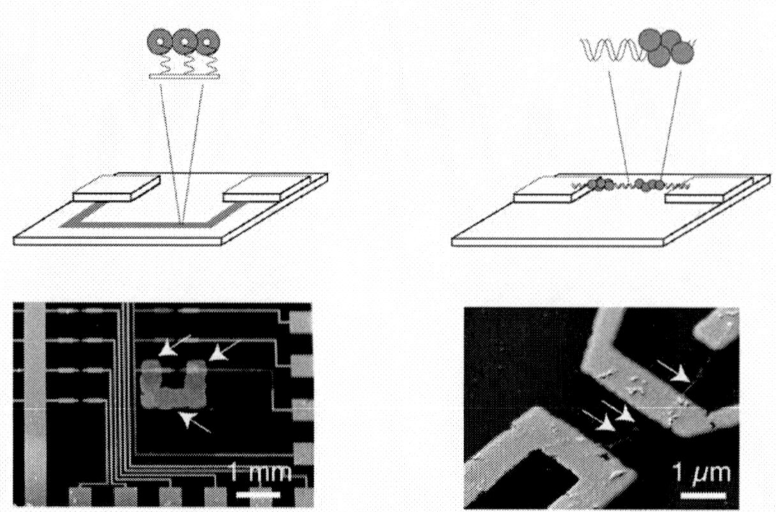

FIGURE 1: Integrated metallic wires based on specific DNA-interactions. Left: Spotted DNA is used to bind the complementary strand that is labeled with a metal seed. This seed is used in an additional enhancement step to grow a metal layer on the DNA-modified regions of the chip (arrows in photo). Right: Individual DNA structures positioned in electrode gaps are utilized as template for the binding of positively charged gold nanoparticles along the DNA (arrows).

Fig. 1 (left) shows a possible approach using surface regions modified with single-stranded DNA (comparable to a DNA chip) that are used to bind complementary DNA that is conjugated to a metal seed (could be a gold nanoparticle or an enzyme that catalyzes metal deposition [1]). This seed is then the starting point for a silver deposition that finally leads to a continuous metal layer in these regions, which is comparable to the microstructured electrodes in its electrical conductivity.

Although this approach uses already molecular principles in coupling the seed to the surface, the first step is still based on spotting techniques so that the resulting structure is still on the micrometer scale. In order to reach the nanoscale, individual molecules should be addressed.

This kind of connection could be realized using defined binding of individual DNA on microstructured substrates, thereby ensuring that the molecule is integrated in the required technical set-up. An example of such a construct is given in Fig. 1 (right), where an individual DNA structure is bridging a microelectrode gap. The DNA was used to bind selectively gold nanoparticles, thereby creating a metal nanowire. Although single-molecule approaches such as optical tweezers allow in principle the manipulation of DNA towards this demonstrated integration, other ways are explored in order to ensure parallel approaches that are compatible with providing significant numbers of single molecule structures for research as well as for future applications.

The approaches for single-molecule positioning onto substrates usually include forces to induce a stretched state of extended DNA molecules. Therefore, a moving meniscus or fluid phase, but also electric fields can be used. These three approaches are reviewed in the following, and examples of their application for integrating DNA into microelectrode setups are discussed.

MOVING MENISCUS

Moving meniscus-based techniques represent an established approach to stretch surface-immobilized DNA [2]. Usually done on functionalized (silanized) glass substrates in order to facilitate high-resolution fluorescence studies of DNA, this method utilizes a moving fluid-gas interface to align molecules that stick (with one end) to the substrate. This moving interface can be realized by sliding coverslips [3], by moving a substrate out of a solution [4] or by using droplets in the process of evaporation [5]. This technique can be applied to microstructured surfaces, and given the right alignment between the receding interface and the underlying structure, an oriented binding can be realized [6].

As in the case of other fluid-based approaches, the local flow pattern in the moment of adsorption is of key importance. Especially for the use of drying droplets (that is potentially open for high throughput) these local patterns are not easily to predict. Experiments with high DNA concentrations allowed for the mapping of these conditions, using adsorbed DNA that was imaged in fluorescence contrast. This can be done on a microscopic scale, as demonstrated for the gap in the figure in Table 1 (receding meniscus, high concentration). Moreover, even for droplets covering substantial parts of a half-inch chip, the local fluidic behavior can be mapped by studying the DNA adsorption in the individual gaps.

	receding meniscus (microstructure guided) 0.001-300 ng/µl 0.1-50 µl		fluid flow (flow chamber) 250 ng/ml, 1ml/min	dielectrophoresis in electrode gap 3-300 ng/µl, 1 V/µm, 1 MHz
	high concentration	low concentration		
bridged gaps	>50%	15-20%	20%	-
advantages	efficient	single molecules	single molecules	confined to gap
limitations	often DNA bundles	low efficiency	low efficiency	low efficiency

TABLE 1. Comparison of different approaches for positioning of DNA by parallel means

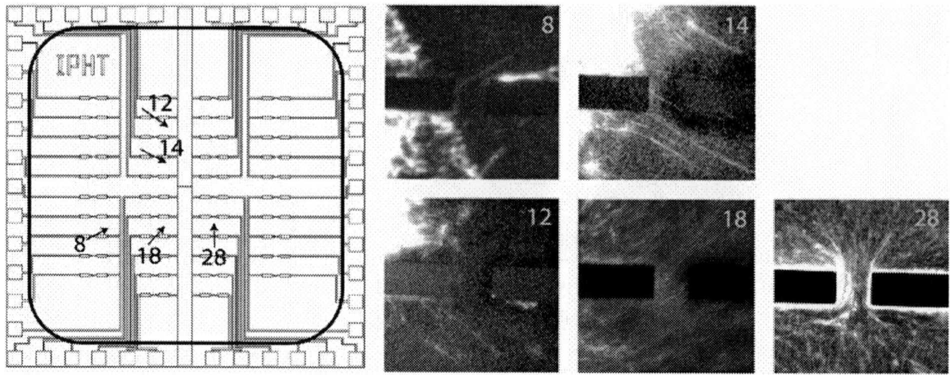

FIGURE 2. Using adsorption of labeled DNA to reveal the local adsorption behavior from a drying droplet onto a microelectrode array. Left: Schematics of the chip with the outline of an ideal droplet and the orientation of adsorbed DNA for several numbered locations (cf. fluorescence micrographs, right) over the chip. Right: Selection of fluorescence micrographs revealing the local directional adsorption

Such an approach is demonstrated in Fig. 2 using fluorescently labeled long DNA (Lambda-phage DNA) that is adsorbed onto a silicon oxide chip substrate. The DNA solution was applied as droplet (see outline, Fig. 2 left), and the drying process lead to the directed adsorption of DNA onto the substrate surface. Fluorescence microscopy revealed the orientation of the adsorbed DNA that reflects the fluid movement during the adsorption process. The imaged DNA molecules show a preferred orientation (Fig. 2, right) that can be utilized to map the local fluid movement (arrows in Fig. 2, left, correspond with the images). In the case of gap (and image) no. 28, only a low DNA-binding onto the electrode structures (gold) becomes visible. This points to an early dewetting of these areas, so that the fluid flow is guided (and thereby concentrated) through the gap, resulting in the observed enhanced DNA adsorption there.

POSITIONING BY FLUID FLOW

The application of directed fluid flow containing DNA molecules onto a substrate with certain affinity for DNA binding represents another approach for DNA positioning with potential for parallelization. Comparable to the receding meniscus discussed before, the molecules should bind on one end prior to be forced to be stretched in a given direction. This stretching was induced by the receding meniscus there, but can also be realized using fluid flow. Although requiring a more sophisticated setup, the fluid flow approach is capable of ensuring a highly defined flow pattern over the substrate surface, as visualized by DNA adsorption on mica [7]. As already mentioned for the meniscus approach, this aligned adsorption can be superimposed onto microstructured substrates. Such substrates can provide the needed connection to a macroscopic technical environment using e.g. microelectrodes in order to realize a true integration of the molecular structures. The demonstration of this approach using microfluidic flow chambers yielded electrode gaps with individual double-stranded DNA molecules in gap structures prepared in a parallel manner [7].

22

However, the demonstration was hampered by a low efficiency of bridged gaps, and the required sophisticated setup compared to the droplet-meniscus approach.

DIELECTROPHORESIS

The described combination of microstructured substrates and fluid flow or moving meniscus-based positioning approaches have the potential for the realization of structures and devices that integrate molecular and microtechnical components in a defined way in a parallel manner. They allow the preparation of a large number of aligned molecular structures on one substrate in one step. However, in order to realize different structures (e.g., various DNA sequences or lengths) at defined locations, these areas have to be defined before the DNA binding (e.g. by different DNA-binding molecules, such as short DNA oligonucleotides). It would require a nanolocal (bio) chemical modification. Such modifications could be realized e.g. by nanospotting that is quite difficult for the needed precisions in the lower micrometer range. Another and more promising option is the use of electrical fields to position the structures onto microelectrode arrays. Electrical interactions were used to attach small structures onto substrates, such as micro- and nanoparticles [8] [9] [10] or DNA [11] [12] [13]. In the case of elongated structures such as long DNA, DC-fields (electrophoresis) are usually applied to locate the molecules at a certain area, but AC (dielectrophoresis) to induce a stretched formation in between electrodes. The latter approach thereby allows a controlled alignment of the stretched molecules and even the wiring of their ends that are pinned down to electrodes. The advantage of this technique is the limitation of adsorption to the chosen electrode gap, minimizing unspecific binding effectively. However, limiting the manipulation to just one individual molecule is not an easy task, leaving us usually with bundles of DNA in the gap. Dielectrophoresis represents a method based on a complex mechanism that requires extensive experience and knowledge in order to control the results – but has an outstanding potential for manipulation on the single molecule scale as impressively demonstrated by Washizu [13] [14] [15].

CONCLUSION

Techniques for a parallel manipulation of individual nanoscale structures and their defined immobilization are available and utilized in research laboratories. Whereas the fluid-based stretching methods (fluid flow as well as moving meniscus) provide an established method for highly parallel and – in connection with microstructured substrates – precise positioning of individual molecular structures, the electrical field-based methods enable highly selective (and variable) manipulation capabilities with the need for prestructured electrodes and the requirement of a certain understanding of the involved complex processes.

ACKNOWLEDGMENTS

We thank M. Köhler for his contributions to the establishment of this field at the IPHT and F. Jahn for SEM imaging.

Funding by the European Commission (6th Framework Programme, STRP 013775), and the Volkswagen Foundation (I/79 311, I/80 070) is gratefully acknowledged.

REFERENCES

1. R. Möller, R. D. Powell, J. F. Hainfeld, and W. Fritzsche, *Nano Letters* **5** (7), 1475 (2005).
2. A. Bensimon, A. Simon, A. Chiffaudel, V. Croquette, F. Heslot, and D. Bensimon, *Science* **265**, 2096 (1994).
3. H. Yokota, F. Johnson, H. Lu, R. M. Robinson, A. M. Belu, M. D. Garrison, B. D. Ratner, B. J. Trask, and D. L. Miller, *Nucleic Acids Res* **25** (5), 1064 (1997).
4. K. Otobe and T. Ohtani, *Nucleic Acids Res* **29** (22), E109 (2001).
5. J. Jing, J. Reed et al., *Proc Natl Acad Sci U S A* **95** (14), 8046 (1998).
6. G. Maubach and W. Fritzsche, *Nano Letters* **4** (4), 607 (2004).
7. G. Maubach, A. Csaki, R. Seidel, M. Mertig, W. Pompe, D. Born, and W. Fritzsche, *Nanotechnology* **14**, 546 (2003).
8. R. C. Hayward, D. A. Saville, and I. A. Aksay, *Nature* **404**, 56 (2000).
9. R. Kretschmer and W. Fritzsche, *Langmuir* **20**, 11797 (2004).
10. C. R. Barry, J. Gu, and H. O. Jacobs, *Nano Lett* **5** (10), 2078 (2005).
11. R. Hölzel and F. F. Bier, *IEE Proc Nanobiotechnol* **150** (2), 47 (2003).
12. R. G. Sosnowski, E. Tu, W. F. Butler, J. P. O'Connell, and M. J. Heller, *Proceedings of the National Academy of Science of the USA* **94**, 1119 (1997).
13. M. Washizu and O. Kurosawa, *IEEE Transactions of Industrial Applications* **26** (6), 1165 (1990).
14. H. Kabata, O. Kurosawa, I. Arai, M. Washizu, S. A. Margarson, R. E. Glass, and N. Shimamoto, *Science* **262**, 1561 (1993).
15. H. Kabata, W. Okada, and M. Washizu, *Japanes Journal of Applied Physics* **39**, 7164 (2000).

Rolling Circle Amplification For Spatially Directed Synthesis Of A Solid Phase Anchored Single-Stranded DNA Molecule

Edda Reiß, Ralph Hölzel, Markus von Nickisch-Rosenegk
and Frank F. Bier

Fraunhofer Institute for Biomedical Engineering
Department of Molecular Bioanalytics & Bioelectronics
Arthur-Scheunert-Allee 114-116, 14558 Nuthetal, Germany

Abstract. In this article the usefulness of the enzyme phi29 DNA polymerase and the principle of rolling circle amplification (RCA) for creating single-stranded DNA (ssDNA) nanostructures is described. Currently we are working on the spatial orientation of a growing ssDNA molecule during its RCA-based synthesis by the application of a hydrodynamic force. Starting at an immobilized primer at single molecule level, the aim is to construct a nanostructure of known location and orientation, providing multiple repeating binding sites that can be addressed via complementary base-pairing. Proof-of-principle experiments demonstrate the potential of the enzymatic reaction. ssDNA molecules of more than 20 μm length were created at an immobilized primer and detected by means of fluorescence microscopy.

Keywords: DNA, phi29 DNA polymerase, rolling circle amplification, CircLigase™
PACS: 87.14.Gg, 81.07.–b

INTRODUCTION

DNA self-assembly via complementary base pairing is a promising tool for the arrangement of nanostructures with defined shapes and has been studied intensively during the last years [1].

In this context the enzyme phi29 DNA polymerase and the principle of RCA could be a useful means in the nanoscale construction kit. This enzyme combines a high processivity with very good strand-displacement properties and incorporates at least 70.000 nucleotides per binding event under isothermal conditions [2]. RCA using phi29 DNA polymerase (or other polymerases with strand-displacement properties) has been widely used for on-chip signal amplification reactions [3, 4, 5]. In these approaches a primer is attached to a surface directly or indirectly (e.g. coupled to an antibody) with its 5'-end. Next, a circular ssDNA molecule is hybridized to the free 3'-end of the primer and RCA results in a long single-stranded concatemeric product which is composed of repeating units of a sequence that is complementary to the sequence of the circular ssDNA molecule. In this way, multiple hybridization sites for detector-oligonucleotides are created, resulting in a strong signal amplification in comparison to a single hybridization site.

The usefulness of such a long concatemeric ssDNA sequence produced by RCA as a nanotemplate has already been demonstrated by Beyer et al. [6].

Or aim is to carry out the RCA under the influence of an external (e.g. hydrodynamic) force. Starting at an immobilized primer, the growing molecule will be spatially directed during its synthesis. As a future perspective, we plan to recombinantly introduce functional DNA sequences (e.g. aptamers) into the concatemeric construct, providing binding sites for adequate ligands.

In this work we demonstrate the potential of this approach by presenting the results of proof-of-principle experiments in which ssDNA molecules of more than 20 µm length were created at an immobilized primer.

MATERIALS & METHODS

Preparation Of The Single-Stranded Circular Template

DNA amplification using phi29 DNA polymerase and the principle of RCA requires a ssDNA template that is circular. For proof-of-principle experiments, the synthetic deoxyoligonucleotide sequence 5'-phosphate-CCTGCGTCGTTTAAGGA AGTACGTGGAAAGTGGCAATCGTGAAGGGACGAATACAAAGGCTACACG GGTCCGGTCATAAAGCGATAAG-3' (Thermo Electron Corporation), termed F1F9F10F5 below (for this and all other sequences mentioned in this article, please refer to reference [7]), was circularized using the enzyme CircLigase™ ssDNA ligase (Epicentre® Biotechnologies) according to the manufacturer's instructions. Under the consumption of ATP, this enzyme catalyzes the intramolecular ligation (e.g. circularization) of ssDNA templates which have a 5'phosphate and a 3'-hydroxyl group. Residual linear ssDNA template was removed by digestion with exonuclease I (Fermentas Life Sciences). The reaction mixture was purified using the QIAquick® Nucleotide Removal Kit (Qiagen). The result of the reaction was controlled by denaturing gel electrophoresis on a 15 % polyacrylamide / 7 M urea gel.

Rolling Circle Amplification

Slide Preparation

After establishing the RCA using the circularized F1F9F10F5 in the liquid phase, the reaction was transferred to a solid support. For that purpose a primer being complementary to a 22 nucleotides long section of the 88 nucleotides long circle was covalently immobilized at an epoxy-modified glass slide.

Briefly, the deoxyoligonucleotide termed cF5 with the sequence 5'-NH$_2$C$_{12}$-CTTATCGCTTTATGACCGGACC-3' (Thermo Electron Corporation) was diluted to 15 µM in 150 mM phosphate buffer (150 mM Na$_2$HPO$_2$/150 mM NaH$_2$PO$_4$) at pH 8. The solution was deposited onto epoxide slides (Corning Incorporated) in an array format using the microarray spotter Sciflexarrayer (Scienion AG). After the spotting process, the slides were incubated for a period of two days at approximately 76 % relative humidity at room temperature and then washed as follows: five minutes in 0,1 % Triton X-100, twice two minutes in 1 mM HCl, ten minutes in 100 mM KCl, and two minutes in de-ionized H_2O. The blocking step was carried out immediately afterwards by incubating the slides for 15 minutes in 50 mM ethanolamine with 0,1 % sodiumdodecylsulfate (SDS) in 100 mM Tris-Cl, pH 9 at 50 °C. Next, the slides were washed for two minutes in de-ionized water and dried by centrifugation. Slides were stored vacuum-packed with a desiccant until used.

Rolling Circle Amplification At An Immobilized Primer

Reaction chambers of 7 x 7 mm were created on the slide surface using a ProPlate™ multi-array slide module (Grace Bio-Labs Inc.). The circularized template F1F9F10F5 was diluted to 10 nM in 3 x SSC (1 x SSC: 0,15 M NaCl, 0,015 M sodium citrate, pH 7) with 0,1 % SDS and the solution kept at 80 °C for five minutes. In the negative control chamber, the template was omitted from the buffer solution. 50 µL of the hot solution were applied per reaction chamber and sealed with an adhesive foil. Hybridization was carried out for one hour at 30 °C. The reaction mixture was removed and the chambers were washed consecutively with 120 µL of 2 x SSC with 0,2 % SDS, 2 x SSC and 0,2 x SSC for 5 minutes each at room temperature. Next, 50 µL of phi29 DNA polymerase reaction mixture was added to each chamber. The reaction mixture contained 0,5 U/µL phi29 DNA polymerase (Fermentas Life Sciences), 1 mM of each dNTP (Bioline) and 0,2 mg mL^{-1} bovine serum albumin (New England Biolabs) in 1 x reaction buffer supplied with the enzyme. The reaction mixture was incubated at 30 °C for 18 hours.

Fluorescence Detection Of The RCA Product

Fluorescence-Labeling Of RCA Product By Hybridization

The phi29 DNA polymerase reaction mixture was removed and the chambers were washed for five minutes with 120 µL 2 x SSC with 0,2 % SDS. Then 80 µL of 500 nM 5'Cy3-GGTCCGGTCATAAAGCGATAAG-3' (Thermo Electron Corporation), termed 5'Cy3F5 below, in 3 x SSC with 0,1 % SDS was added to each chamber and hybridization was carried out for one hour at 30 °C. The hybridization mixture was removed and the chambers were washed as described above. The slide was dried under a stream of nitrogen.

Image Acquisition

Fluorescence images were taken with an epifluorescence microscope (Leica DMRB) equipped with a CCD camera (Kappa PS 2 C) and appropriate software control. A Leica PL FLUOTAR objective (40x, NA = 0,7) was used and a Leica N2.1 filter cube (green excitation range) was applied in the light path. Slides were covered with a standard cover glass (Carl Roth) using a drop of bidistilled water as mounting medium.

RESULTS & DISCUSSION

RCA requires a circular ssDNA template. Therefore, enzymatic circularization of the phosphorylated sequence F1F9F10F5 using the enzyme CircLigase was established. Figure 1 shows the result of the reaction on a denaturing polyacrylamide gel. In lane 1, an aliquot of the reaction mixture without CircLigase (negative control) was loaded onto the gel, showing the mobility of the linear ssDNA form. Lane 2 shows the result of the enzymatic circularization of the linear template using CircLigase ssDNA ligase: an additional band appears which migrates retarded through the gel matrix. This is due to the enzyme-catalyzed conformational change of a part of the deoxyoligonucleotide molecules from linear to circular form. Lane 3 shows the reaction product after exonuclease I treatment. The faster moving, linear form has disappeared, whereas the slower moving form has not been digested by exonuclease I, proving that it is indeed circular.

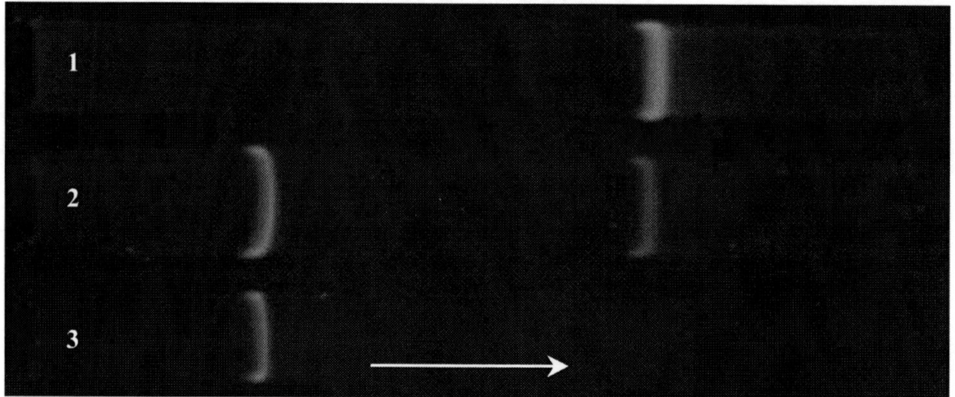

FIGURE 1. Documentation of the circularization reaction on a 15 % polyacrylamide / 7 M urea gel which was post-stained with $1 : 10^4$ diluted Sybr Green II in 1 x TBE (89 mM Tris-borate, 2 mM EDTA, pH 8.3). Approximately 3 pmol ssDNA were applied per lane. Lane 1: negative control (CircLigase omitted), lane 2: product after CircLigase reaction, lane 3: product of lane 2 followed by digestion of linear template by exonuclease I. The arrow indicates the running direction.

The cicular ssDNA form of the sequence F1F9F10F5 created in this way was then applied as a template in liquid phase RCA using phi29 DNA polymerase. Primed circular F1F9F10F5 as starting material resulted in a high molecular weight product band, whereas the same amount of primed linear template (negative control) gave no product band at all (confirmed via alkaline agarose gel electrophoresis, results not shown).

Finally, RCA was carried out starting at a surface-bound primer. The amplification product was labeled by hybridization with a complementary oligonucleotide carrying a Cy3-tag. Spots were visualized microscopically. Figure 2 shows a typical result of the reaction. The left side shows a spot after RCA with the circular template F1F9F10F5 at the immobilized primer cF5. It can be clearly seen that long DNA molecules (> 20 μm) are the product of the reaction, giving the spot an irregular shape. In contrast to this, a control reaction without circular template leads to a smooth shape of the spot after hybridization with 5'Cy3F5 (Fig. 2, right side). Furthermore, signal intensity for the negative control was much weaker than for the positive reaction, as multiple hybridization sites for the detector-oligonucleotide 5'Cy3F5 were created by the RCA reaction.

Immobilizing a primer (5'-NH2-C12-GTACTTCCTTAAACGACGCAGG-3', termed cF1) that was complementary to another 22 nucleotides long section of the circular F1F9F10F5, but not to 5'Cy3F5, which was used for detection, led to the same results for the positive reaction, whereas no hybridization signal was detected for the negative control (results not shown).

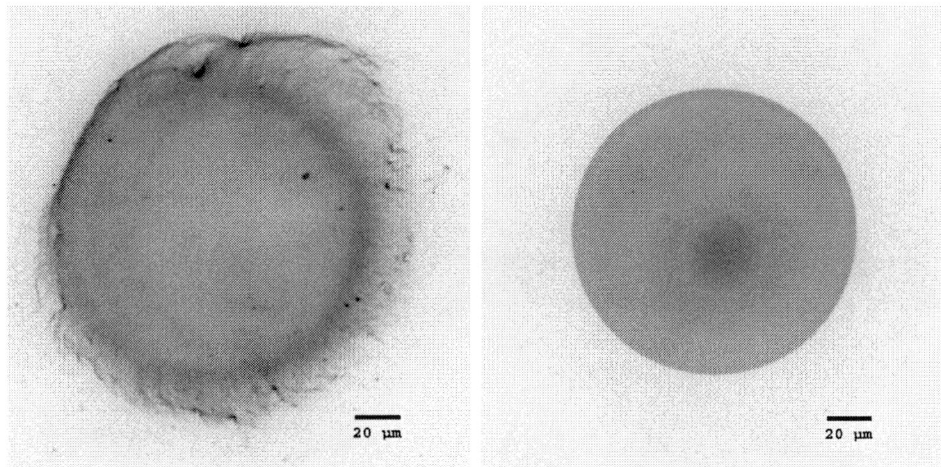

FIGURE 2. Negatives of fluorescence images of spots after RCA reaction at an immobilized primer and labeling with a complementary Cy3-tagged oligonucleotide sequence. Left image: circular template was hybridized to the immobilized primer, and the RCA reaction resulted in long (> 20 µm) ssDNA molecules. Right image: negative control, all steps of the reaction were carried out as in left image, but the hybridization step of the circular template was omitted. Hybridization of complementary Cy3-tagged oligonucleotide resulted in a smooth surface appearance. Exposure time for the right image was approximately 8 times longer than that for the left image.

CONCLUSIONS

The findings made in these preliminary experiments demonstrate that RCA using phi29 DNA polymerase is a promising tool for creating DNA based nanostructures. The subject of our further efforts is the spatial orientation of the enzymatic reaction using a hydrodynamic force and subsequently scaling down the spatial resolution to a single molecule level using Dip Pen Nanolithographie™.

ACKNOWLEDGMENTS

This work was supported by the European Union's sixth framework programme under the project name NUCAN (Nucleic Acid Based Nanostructures), contract no. STREP 013775. In this context, the authors would like to thank Christof Niemeyer and

Kersten Rabe from the University of Dortmund for communicating sequence information. Furthermore, the authors are indebted to Heiko Andresen for helpful discussions.

REFERENCES

1. K. V. Gothelf and T. H. LaBean, *Org. Biomolecul. Chem.* **3**, 4023-4037 (2005).
2. L. Blanco, A. Bernad, J. M. Lázaro, G. Martín, C. Garmendia and M. Salas, *J. Biol. Chem.* **264**, 8935-8940 (1989).
3. P. M. Lizardi, X. Huang, Z. Zhu, P. Bray-Ward, D. C. Thomas and D. C. Ward, *Nat. Genet.* **19**, 225-232 (1998).
4. G. Nallur, C. Luo, L. Fang, S. Cooley, V. Dave, J. Lambert, K. Kukanskis, S. Kingsmore, R. Lasken and B. Schweitzer, *Nucleic Acids Res.* **29**, (2001).
5. B. Schweitzer, S. Wiltshire, J. Lambert, S. O'Malley, K. Kukanskis, Z. Zhu, S. F. Kingsmore, P. M. Lizardi and D. C. Ward, *Proc. Natl. Acad. Sci. U.S.A.* **97**, 10113-10119 (2000).
6. S. Beyer, P. Nickels and F. C. Simmel, *Nano Lett.* **5**, 719-722 (2005).
7. U. Feldkamp, H. Schroeder and C. M. Niemeyer, *J. Biomol. Struct. Dyn.* **23**, 657-666 (2006).

A Novel Approach To Self-organized Pattern Formation Of Biomolecules At Silicon Surfaces

A.Wolff, W. Fritzsche, M. Kittler*, X. Yu*, M. Reiche#, T. Wilhelm#, M. Seibt+ and O. Voß+

Institute for Physical High Technology, Albert-Einstein-Strasse 9, 07745 Jena, Germany
**IHP, Im Technologiepark 25, 15236 Frankfurt (Oder), Germany*
IHP/BTU Joint Lab, Konrad-Wachsmann-Allee 1,03046 Cottbus, Germany
 # MPI f. Mikrostrukturphysik, Weinberg 2, 06120 Halle, Germany
+Georg-August-Univ. Göttingen, Friedrich-Hund-Platz 1, 37077 Germany

Abstract. A new approach for the placement of biomolecules at Si surfaces is to utilize dislocation networks formed by direct wafer bonding. They provide the option for the interaction of biomolecules with these charged dislocation networks, resulting in patterned immobilization of the molecules. This would allow a combination of Si electronics with biological applications. Such parallel approaches open novel ways for the study of phenomena like DNA conductivity and the characterization of metalized DNA or DNA-conjugated nanoparticles.

Keywords: silicon wafer, dislocation networks, DNA
PACS: 61.72.Lk, 82.39.Pj

INTRODUCTION

Organization and defined placement of biomolecules - as useful building blocks for nanoelectronics - at Si surfaces is a critical issue for successful combinations of Si electronics with biological applications and for large-scale integration. Using Coulomb interactions of biomolecules with dislocations in silicon could overcome this problem. Dislocations will form negatively (positively) charged lines in n-type (p-type) Si connected with an electrostatic barrier. A space charge region (Read cylinder) is surrounding these lines [1], [2], [3].

A regular dislocation network can be formed at the interface of misoriented wafers with electric fields, formed by the line charges, reaching up to a few kV/cm [4]. Misorientation during the direct bonding process with tunable rotation angles (twist) allows the variation of the distance between the dislocations in a range from about 10 nm to a few microns. This size range matches with the extent of protein or DNA molecules and their applications in nanobiotechnology. The minimum distance between the dislocations is about 20 nm. EBIC technique (Electron Beam Induced Current) showed the existence of a distinct electrostatic barrier/field at the interface.

CP 859, *DNA-Based Nanoscale Integration: International Symposium,* edited by W. Fritzsche

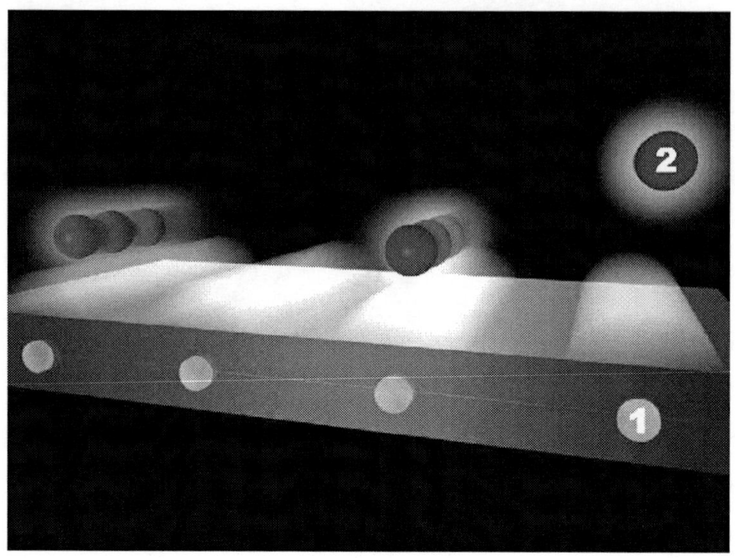

FIGURE 1. Schematic illustration of the principle of molecular pattern formation at bonded interfaces: Dislocation lines (1) induce electrical fields. Biomolecules (2) are attracted to these fields due to Coloumb interactions.

Preparation of Substrates

Grain boundaries were generated by direct wafer bonding [5],[6] and subsequent annealing of differently doped, (100) oriented, H-terminated (hydrophobic) Si crystals. To adjust mesh widths in the range of 5-220 nm it is possible to change the rotation (twist) angles between the Si crystals in the range of 0.1° to 4.5° during the bonding process. For smaller distances angles up to 26.5° have been chosen to receive a width of 2.5 nm minimum. Quality control was done by IR imaging.

Molecular dynamics simulation considered the parameters and angles for misalignment. Important parameters for the electric field of dislocation networks are currently being optimized and will allow the producing of patterns of the electric stray field above the Si surfaces that will be strong enough for a self-organized assembly of biomolecules on top of the samples.

To assure the interaction of molecules from solution with the electric field of the dislocation lines one side of the bonded wafers has to be thinned down. Multiple oxidation-etch-cycles thinned down the wafer to 50-260 nm so that the grain boundary is located close to the crystal surface [7]. For very thin top layers this method yields irregular interfaces and additional defects. The structural characterization of the grain boundaries is performed with TEM, generally using the diffraction contrast. The thickness of Si layers, the smoothness of the grain boundary and possible oxygen-related precipitates at the interface, were investigated using TEM cross-section images.

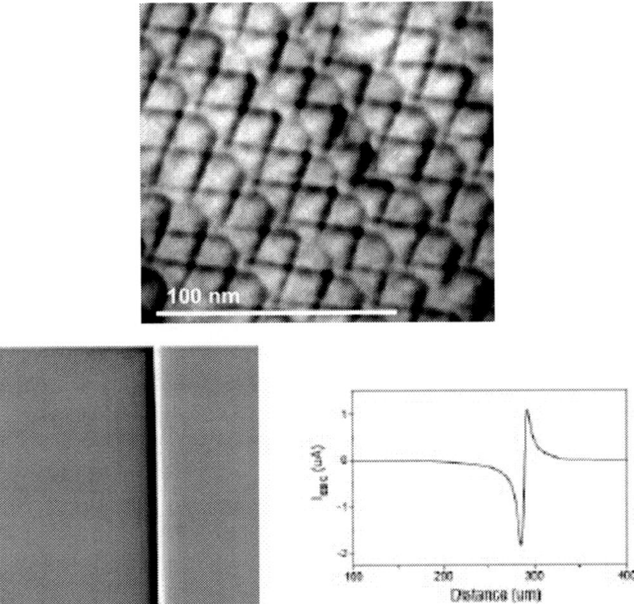

FIGURE 2. TEM image of a screw dislocation network (top) and an image of an GB-EBIC measurement of a p-type sample (bottom).

Grain Boundary-Electron Beam Induced Current measurements reveal potential barriers for the bonded interfaces of about 60-100 meV. For p-type as-bonded Si this barrier is higher than for n-type samples. An equal current direction for both samples could be observed. Different scanning force microscope techniques such as scanning capacitance and electrical force microprobe could show the presence of an electrical field of 50-120 meV at an distance of 50nm from the surface.

Another strategy to increase the dislocation line charge for applications from solution could be the controlled decoration of the dislocations with metal atoms. A strong increase of the charge is expected. Suitable impurities are in testing progress. Gold seems to be applicable for a homogeneous decoration [8],[9]. It has two deep levels in crystalline silicon, i.e. an acceptor state at EC -0.55 eV and a donor level at E_V +0.34 eV. It is also well soluble. High gold concentrations results in a substitutional collection by dislocation strain fields. In n-type samples the acceptor level is pinned to the conduction band, in p-type the donor level is pinned to the valence band.

Biomolecules

Compatible biomolecules for sensing the electrical field of the generated networks and for further wiring and electrical applications are DNA strands. Because of their capability of specific complementary recognition it is possible to use these molecules for a controlled setup of building blocks and their subsequent combination into defined nanostructures. The binding of metal ions and the deposition of metal are further possibilities for a functionalization towards nanoelectronics. A widely established technology with a comprehensive toolbox of enzymes allows the molecular tailoring of desired DNA strands. Last but not least the mechanical properties and rigidity of DNA molecules are suitable for large-scale integration. Two species of DNA molecules will be tested for the application on the generated fields: Long DNA such as DNA of the phage lambda (lambda DNA) and so called G-wire DNA.

Lambda DNA is a 48,5 kb double strand with a length of about 16 μm. Due to the application of many restriction enzymes it is possible to manipulate the DNA strand. So one can hybridize the long strand with its single-stranded (ss) overhang at the ends (cos sequences) to surface bound, complementary ssDNA. This terminal binding can then be used for stretching by application of forces due to a moving meniscus or a fluid flow (Fig. 3, left). Further metallization steps could be performed on these strands [10],[11]. For immobilization – e. g. for AFM imaging, long double-stranded (ds) DNA is usually adsorbed onto mica. This crystalline substrate does not lead to ordered absorption of DNA, probably due to the high flexibility of the biomolecule and/or the weak interactions of the crystal. Therefore, defined arrangement of DNA on surface requires microstructures.

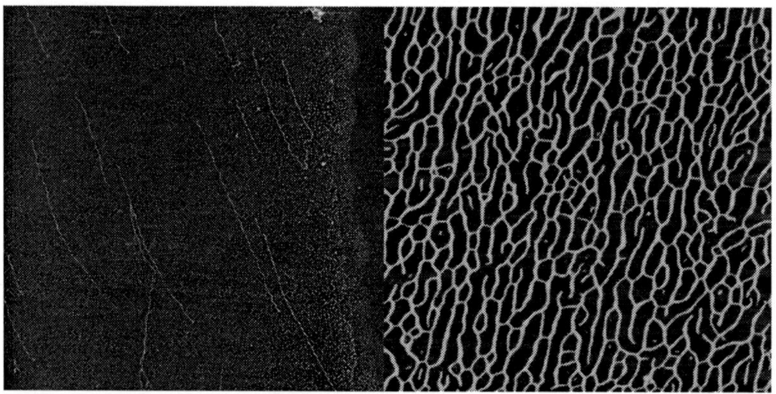

FIGURE 3. AFM images of DNA structures on mica. Left: stretched lambda DNA. Right: G-wire network.

In contrast to dsDNA, G-wires are stiffer. These superstructures self assemble in a highly specific way through Guanine (G) quartets formed by guanine-rich DNA oligonucleotides. The resulting assembled extended DNA structures exhibit an alignment along crystallographic orientations on mica surfaces [12],[13]. Highly concentrated solutions assemble into networks on mica surfaces (Fig. 3, right). A functionalization, for example with biotin (as a standard coupling unit), is also possible. It provides an interesting approach to ordered and positioned molecular arrangements based on small units.

We study the adsorption behavior of dsDNA and G-wires onto mica in order to understand the underlying mechanism. This understanding could lead us to the controlled positioning of biomolecules onto substrates with dislocation networks.

ACKNOWLEDGMENTS

This work is supported by the Volkswagen Stiftung. VW Project: I/80 070.

REFERENCES

1. W.T. Read, *Phil Mag* **45**, 775 (1954) and **45**, 1119 (1954).
2. H. Alexander, "Versetzungen in Halbleitern" in "Festkörperphysik für die Informationstechnik", Jülich 1999, pp. 10.1 – 10.15.
3. W. Schröter, H. Cerva, "Interaction of Point Defects with Dislocations in Si and Ge: Electrical and Optical Effects" in *Defect Interaction and Clustering in Semiconductors,* Ed. S. Pizzini, Solid State Phenomena,2002, pp. 85-86.
4. S. Pandelov, W. Seifert, M. Kittler, J. Reif, *J. Phys.: Condens. Matter* **14,** 13161 (2001)
5. J.B. Lasky, S.R. Stiffler, F.R. White, J.R. Albernathy, Silicon-on-Insulator (SOI) by Bonding and Etch-Back, Proc. of the IEEE Internat. Electronic Device Meeting, IEEE, Piscataway, NJ, (1985) 684. J.B. Lasky, Appl. *Phys. Lett.* **48**, 78 (1986).
6. M. Shimbo, K. Furukawa, K. Fukuda, K. Tanzawa, J. *Appl. Phys.* **60**, 2987 (1986).
7. B. Aspar, E. Jalaguier, A. Mas, C. Locatelli, O. Rayssac, H. Moriceau, S. Pocas, A. M. Papon, J. F. Michaud, M. Bruel, *Electron. Lett.* **35**, 1024 (1999). B. Aspar, M. Bruel, H. Moriceau, C. Maleville, T. Poumeyrol, A. M. Papon, A. Claverie, G. Benassayag, A. J. Auberton-Herve, T. Barge, *Microelectron. Eng.* **36**, 233 (1997). M. Bruel, *Electron. Lett.* **31**, 1201 (1995).
8. M. Kittler, V. Higgs, C. Ulhaq-Bouillet, *J. Appl. Phys.* **78**, 4573 (1995). V. Kveder, M. Kittler, W. Schröter, *Phys. Rev. B* **63**, 115208 (2001). K. Knobloch, M. Kittler, W. Seifert, *J. Appl. Phys.* **93**, 1069 (2003).
9. O. Voß, Diploma Thesis, Universität Göttingen 2003.
10. G. Maubach and W. Fritzsche, *Nano Letters* **4**, 607-611 (2004).
11. G. Maubach, D. Born, A. Csaki and W. Fritzsche, *Small* **1**, 619-624 (2005).
12. T.C. Marsh, J. Vesenka, E Henderson, *Nucleic Acids Research* **23**, 696-700 (1995).
13. A. Sondermann, C. Holste, R. Möller and W. Fritzsche, in *DNA-Based Molecular Construction* edited by W. Fritzsche, AIP Conference Proceedings 640, 2002, pp. 103.

METALLIZATION

Synthesis of Palladium Conductive DNA-based Nanowires

Khoa Nguyen, Stéphane Streiff, Sébastien Lyonnais
Laurence Goux-Capes, Arianna Filoramo, Marcelo Goffman and
Jean Philippe Bourgoin
Laboratoire d'Electronique Moléculaire SPEC-DRECAM - CEA Saclay

Abstract. We present here a simple method to metallize DNA by Electroless Plating of palladium, a trusty metal for contacting SWNT devices. Indeed, DNA is a promising scaffolding candidate for molecular electronic bottom-up self-assembly approaches of SWNT devices. We report in this work the synthesis and characterization of individual Pd nanowires as thin as 30 nm showing ohmic behavior at room temperature.

Keywords: DNA, Metallization, Nanowires, Electroless Plating
PACS: 81.07.-b; 85.40.Ls; 87.14.Gg

INTRODUCTION

Nanotechnology aims to produce and manipulate well-defined structures at the nanoscale level with high accuracy. However, nowadays is well admitted that conventional technologies based on the "top-down" approach are foreseen to experiment difficulties due to various physical effects that do not scale properly, and most important, related to the cost issues of the fabrication processes at the nanoscale dimension. This is a prompt for the study of new methodologies based on bottom-up approaches. In this framework, DNA molecule is of particular interest since, thanks to its unique intra- and intermolecular recognition properties, it has been already used to build-up nanostructures[1,2] or scaffolds for nanoparticles assembly[3-5] and it can be envisioned for the assembly of devices. Our ambition is also to utilize DNA not only as a positioning scaffold for nanodevices but also as a conducting element. In order to ensure good DNA transport properties when this molecule is deposited on dry substrate like in nanodevice configuration we found that the best way is to chemically modify it by metallization.

During the past 10 years numerous methods to metalize DNA scaffolds have been developed[6]. Different metals have been used in the metalization process. For the techniques based on the ion-exchange mechanisms on the DNA backbone we can quote silver[7,8] or copper[9]. Concerning the intercalation mechanism of metal complexes between the DNA bases, the Pd or Pt complexes have been the more extensively studied[10-18].

Here we report a method based on the Electroless Plating of palladium ions which bind to DNA bases and serve as catalysts for further metal growth[15-18]. Our main concern is to develop a technique in which the metallization is the last step of the assembly process and circuits fabrication. This is indeed essential since DNA loses its

CP 859, *DNA-Based Nanoscale Integration: International Symposium,* edited by W. Fritzsche
© 2006 American Institute of Physics 978-0-7354-0357-4/06/$23.00

recognition and assembly properties after the metallization step. Therefore, we studied the experimental conditions to ensure DNA metallization only after its deposition onto the substrate while minimizing any parasitic metallization of the surrounding surface. Then, we connected the so-obtained nanowires by standard lithographic techniques and we performed transport measurements.

Experimental section

<u>Surface treatment:</u>

In order to limit the parasitic metallization of the surface when the DNA strands are metallized, we perform a surface treatment of SiO_2 substrate with HMDS (HexaMethylDiSilazane). Indeed this methylated surface ($-CH_3$) is non reactive towards the metallization process. The protocol of surface treatment can be decomposed in 5 steps as follows:

1. a 2'' wafer of Silicium (1500Å SiO_2) is piranized for one hour then carefully rinsed in milliQ water
2. a step of RIE (Reactive Ion Etching) in a plasma O_2 is then realized to increase reactivity of the surface (i.e. density of silanols)
3. the wafer is then baked for 15 minutes at 170°C for dehydratation, avoiding HMDS to react with water.
4. Finally the substrate is put in a glass chamber with a recipient containing HMDS solution (99,9% Sigma Aldrich)
5. After half an hour the wafer is carefully rinsed with DI water then ULSI acetone to remove any physisorbed layer of TriMethylSilyl from the reaction of HMDS with left water molecules

<u>DNA deposition:</u>

Linear λ-DNA molecule, 48,502 base pairs in length, is stretched and aligned on the HMDS hydrophobic surface by molecular combing technique using the capillary forces applied by a receding meniscus [19,20]. The used experimental conditions are as follows:

λ-DNA (7 ng/μL, Sigma) is prepared in a 10 mM MES buffer solution pH 5.5, 50 mM NaCl. An HMDS silanated silicium surface is immersed in such solution for one minute. Then, the sample is carefully pulled out at constant velocity (100μm/s). This process ensure the alignment of DNA on the HMDS surface by molecular combing and typical results are showed in figure 1.

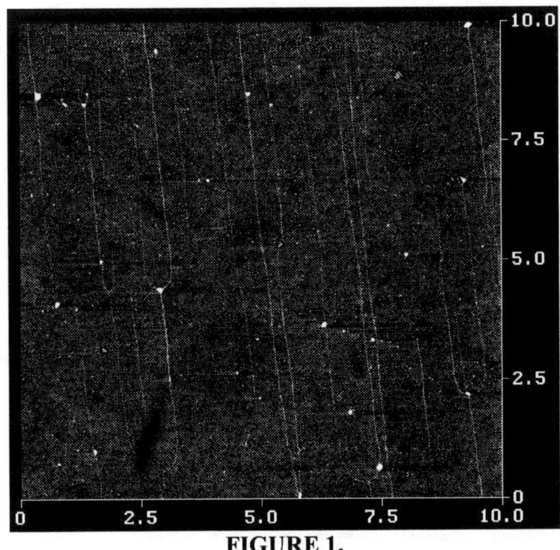

FIGURE 1.
DNA combing on SiO$_2$ surface silanated with HMDS

DNA metallization:

The majority of DNA metallization processes follow three main successive steps [6]. The first step concerns the biomolecule activation: the metal ions or metal complexes bind to DNA, creating reactive metal sites. This activation can take place by an ion exchange mechanism (like for silver as reported in [7,8]) on the DNA backbone or by insertion of the metal complexes between the DNA bases (like in our case and more generally for platinum or palladium complexes [10-18]). In the second step, the reactive metal sites (the bound seeds) are usually treated with a reducing agent. This converts the metal ions or metal complexes in metal nanoclusters fixed on the DNA strand. At the end of these two steps, the DNA strand has some small metal nanoclusters fixed on it, which will successively act as "seeds" for the metallization of the DNA molecules. The third step of the metallization process consists in auto-catalytical growth of the fixed metal seeds on the DNA strand by the addition of new metal ions (or metal complexes) and new reducing agents.

In our experiment, to activate the DNA combed on the HMDS surface the sample is gently immersed in a 10mM palladium chloride solution (24 hours aged) and left there for 22 hours at 25°C. Then, the sample is taken out and a 20 mL drop of 2mM DiMethylAmine Borane (DMAB) pH 7.4 solution was placed onto it and serves as reducing agent. Then, for the clusters growth a 20 ml drop of 2mM palladium solution was injected in the initial drop. Cycling is achieved by tilting the hydrophobic surface removing the resulting drop and placing another drop of palladium then DMAB. The samples were rinsed carefully afterwards with pure water.

Finally metal pads are patterned to anchor and electrically connect one single DNA based nanowire using a standard electron-beam lithography followed by deposition of 1nm Cr and 60 nm Au.

41

Results and Discussion

The dip coating technique used on our hydrophobic surface (contact angle ~72°) yields reproducible results both in terms of density and particularly well stretched DNA of more than 20μm as discussed in the literature [20] (unstretched λ-DNA is 16.1μm). This gives us a basis for evaluating the thickness and yield of our metallization process.

In this work focus we focus our studies on the influence of growth cycles on the nanowires diameter and homogeneity: we keep constant the activation parameters and solution concentrations while varying the number of cycles.

During the first two growth cycles, separated clusters are formed on the DNA template. These clusters will serve as catalytic nucleation centers for the further reduction of palladium resulting in a continuous metal coating after seven growth cycles. We are able to fine-tune the resulted nanowires in terms of average diameter and continuity by changing the number of development steps (as shown in figure 2). Moreover, the synthesized nanowires are of unprecedented homogeneity showing no signs of stochastic branching. This confirms that it is possible to keep the scaffold deposited on the substrate in its geometrical configuration (in this case combed and aligned) while performing the metallization process. Moreover, the Electroless Plating mechanism is extremely well templated by DNA and only few unwanted metallization (parasitic) of the surface is observed. This is due to the effectiveness of the activation step which is long enough (for a palladium concentration of 10mM) to initiate sufficiently close nucleation sites on DNA. The growth cycles are also crucial for the resulting nanowires.

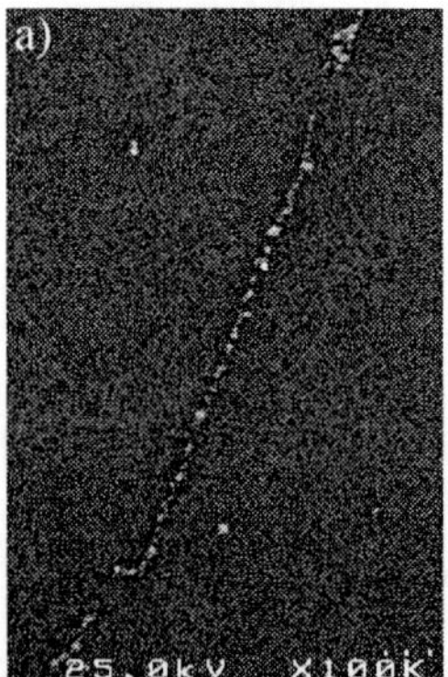

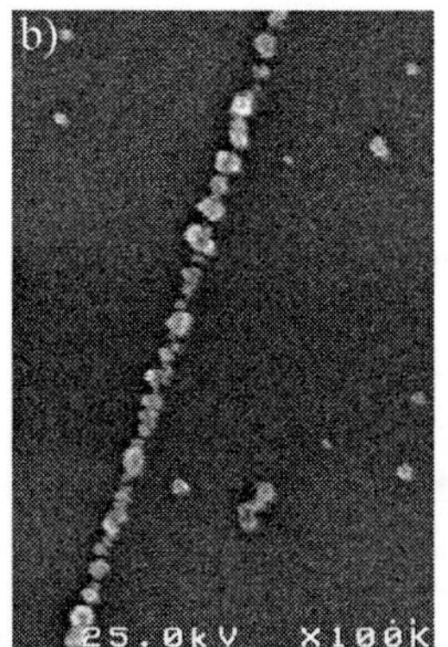

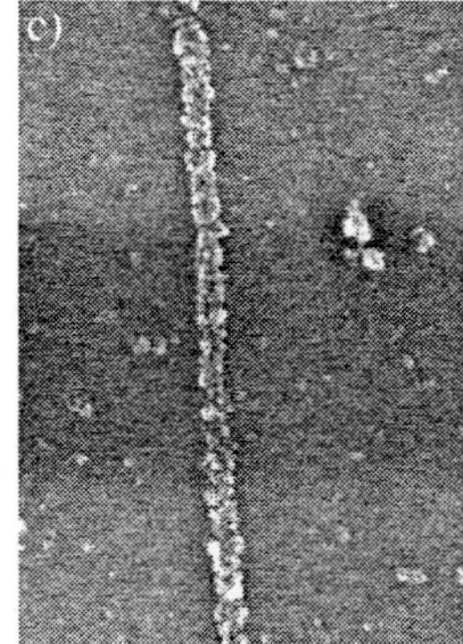

FIGURE 2. Cycling effects on resulted nanowires – Scanning Electron Microscopy images. (a) palladium clusters necklace (b) growth of palladium clusters (c) continuous palladium nanowire – granularity of the wire is to be noted

Thanks to surface passivation with methyls, reagents can be inserted directly on the substrate with little parasitic metallization. Only reagents which are at the immediate proximity of the surface will react due to the extremely fast reduction reaction of palladium complexes. Fast kinetic and controlled injection of the reagents allow the progressive, thus tunable growth of the nanowire from coarse to continuous and extremely regular metallization. It is to be noted that stretching DNA does not destroy its capabilities to template the Electroless Plating of palladium complexes. It may even be favourable to the absence of branching by forcing nucleation centres to be perfectly aligned.

The electrodes for transport measurement are fabricated by standard lithography and evaporation process, exposing metallized DNA to temperature up to 170°C. In spite of this, the nanowire shows no apparent sign of melting as we can see in the SEM images reported in figure 3.a and 3.b recorded respectively before and after electrodes deposition. This point is particularly interesting for electronic applications even if we cannot exclude that the crystal structure may have been changed [18].

We have studied a dozen of nanowires with average diameter ranging from 30 nm up to 60 nm and all show ohmic behavior at room temperature with overall resistance from ~200Ω to 1.5kΩ. These small values for two probes measurements indicate that contact resistance between palladium and gold is very low. We report in figure 3 the results of characterization for a typical nanowire of 60 nm diameter. We measure a resistance of 193Ω and we can extract an upper estimate for the resistivity below 436.10^{-8}Ω.m (neglecting contact resistances) which is about ten times the one of bulk palladium.

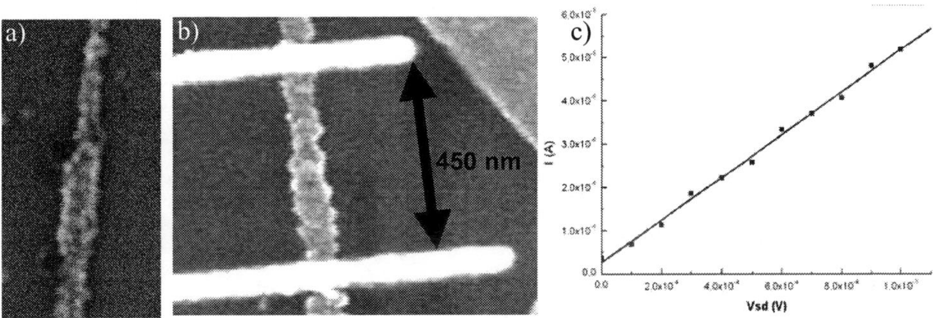

FIGURE 3. Electrical Connections and Measurements
(a) DNA_based palladium nanowire previous to connecting (b) after connection (Au electrodes are 50nm large) (c) Two-probes current–voltage curves of the nanowire (b)

In conclusion, we have demonstrated the synthesis of palladium nanowires templated on a stretched DNA on a surface. The resulting nanowires are tunable in diameter, thin, regular, highly conductive while parasitic metallization is well controlled thanks to the HMDS passivated surface.

These results paves the way for the use of this optimized metallization process to connect devices positioned on a DNA scaffolds in a molecular circuit context.

ACKNOWLEDGMENTS

This work is partially supported by the European project NUCAN – NMP STREP 013775 and by the AC Bio-NT project of French government.

REFERENCES

1. N.C. Seeman, *Nature* **421**, 427-431 (2003).
2. N.C. Seeman, *Nanotechnology* **2**, 149-159 (1991)
3. A.P. Alivisatos, K.P. Johnsson, X.G. Peng, T.E. Wilson, C.J. Loweth, M.P. Bruchez, P.G. Schultz , *Nature* **382**, 609-611 (1996).
4. C.Niemeyer, B.Ceyhan, *Angew. Chem. Int. Ed.* **40**, 3685-3688 (2001)
5. C.A. Mirkin , *Inorg. Chem.* **39**, 2258-2272 (2000)
6. J. Richter, *Physica E* **16**, 157-173 (2003)
7. H. Yan, S.H. Park, G. Finkelstein, J.H. Reif , T.H. LaBean , *Science* **301**,1882-1884 (2003)
8. E.Braun, Y.Eichen, U. Sivan, G. Ben-Joseph, *Nature* **391**, 775-778 (1998)
9. C.F. Monsoon, A.T. Woolley, *Nanoletters* **3**, 359-363 (2003)
10. *Cisplatin: chemistry and biochemistry of a leading anticancer drug*; Lippert, B., Ed.; Wiley-VCH: Weinheim, Germany, 1999.
11. J.P. Macquet, T.Theophanides, Biopolymers **14**, 781-799 (1975)
12. L. Colombi Ciacchi, M.Mertig, R. Seidel, W. Pompe, A. de Vita, *Nanotechnology* **14**, 840-848 (2003)
13. J.P. Macquet, T. Theophanides, Inorg. Chim. Acta, **18**, 189-194 (1976)
14. J.P. Macquet, J.L. Butour, Eur. J. Biochem. **83**, 375-385 (1978)
15. M. Mertig, L. Colombi Ciacchi, R. Seidel, W. Pompe, A. De Vita, *Nanoletters* **2**, 841-844 (2002)
16. J. Richter, M. Mertig , W.Pompe, I. Monch, H. K. Schackert, *Appl. Phys. Lett.* **78**, 536-538 (2001)
17. Z.Deng, C.Mao, *Nanoletters* **3**, 1545-1548 (2003)
18. J. Richter, M. Mertig, W. Pompe, H. Vinzelberg, *Appl. Phys. A* **74**, 725-728 (2002)
19. A. Bensimon, A. Simon, A. Chiffaudel, V. Croquette, F. Heslot, *Science* **265**, 2096-2098 (1994)
20. D. Bensimon, A. Simon, V. Croquette, A. Bensimon, *Phys. Rev. Lett.* **74**, 4754 -4757 (1995)

Directed DNA Metallization: Towards the Construction of Rationally Designed Conductive Nanodevices

Glenn A. Burley,[*] Johannes Gierlich, David M. Hammond, Phillipp M.E., Gramlich and Thomas Carell[*]

Fakultät für Chemie und Pharmazie, Ludwig-Maximilian-Universität München, Butenandtstr. 5-13 (Haus F), D-81377 München, Germany
Email: Glenn.Burley@cup.uni-muenchen.de; Thomas.Carell@cup.uni-muenchen.de

Abstract. Genes of interest can be selectively metallized via the incorporation of modified triphosphates. These triphosphates bear functions that can be further derivatised with aldehyde groups via the use of click chemistry. Treatment of the aldehyde-labeled gene mixture with the Tollens reagent, followed by a development process results in the selective metallization of the gene of interest in the presence of natural DNA strands.

Keywords: Molecular Lithography; Polymerase; DNA; click chemistry; Nanowires; modified triphosphates
PACS: 81.07.-b; 81.16.Dn; 81.16.Fg; 81.16.Rf; 81.20.-n; 81.20.Ka

INTRODUCTION

Nucleic acids have the potential for use as both a generic and a genetic material. Whereas nature exploits the sequence-specific templating properties of nucleic acids for the storage and flow of genetic information, material scientists and engineers are just beginning to use nucleic acids for the construction of nucleic acid-mediated nanostructures. Since the pioneering work by the Seeman group established the guiding principles of DNA-programmed self-assembly, non-biological nucleic acid programmed self-assembly has been used for a variety of applications such as detecting disease states or genetic variation,[1] implementing molecular computation, the construction of nanorobotic devices and learning circuits, and nano-templating of functional molecules.

Our research group is interested in exploiting self-assembling macromolecules as instructional tools for the construction of *rationally* designed, conductive nano devices. DNA in this respect is an outstanding architecture for directing the self-assembly of functional molecules in a defined multi-dimensional matrix. Unfortunately naturally-occurring DNA does not conduct electrical charge at the efficiency that is required for electronic devices. DNA metallization procedures were subsequently developed in order to increase the conductivity of DNA nanostructures, thereby marrying self-assembly and conductivity. Typical metallization processes involve the chemical reduction of DNA-complexed metal salts (e.g., Ag, Pd, Pt, and Cu),[2] resulting in uniformly metallised DNA architectures. Although this method produces DNA-templated conductive wires, global metal deposition of DNA templates prevents its use for the fabrication of DNA-templated electronic components. Therefore a method that allows sequence-selective metallization of DNA is required.[2] In this conference proceedings we summarise our results towards this aim of gene-selective metallisation via the incorporation of modified nucleoside building blocks that can direct metal deposition down to the gene-specific level.

RESULTS AND DISCUSSION

Investigation of click chemistry on alkyne-modified DNA

We devised an alternative labelling strategy based on the copper (I) mediated azide/alkyne [3+2] cycloaddition initially developed by Huisgen and more recently termed "Click Chemistry" by the Sharpless group.[3] Alkyne nucleoside monomers such as **1** and **2** were synthesized according to procedures published elsewhere.[4]

To evaluate whether click chemistry would be a useful post-synthetic method for high density labeling of alkyne-modified DNA, nucleosides **1** and **2** were incorporated into a series of 16-mer ODNs via their corresponding phosphoramidites (Figure 1).

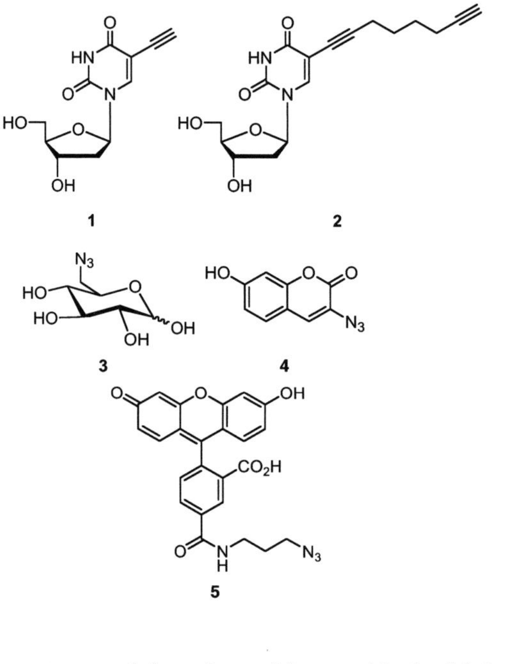

FIGURE 1.

In order to circumvent potential steric problems with the high density labeling of ODNs containing the alkyne **1**, we also prepared the nucleoside **2** comprising a sterically less demanding terminal alkyne. The alkyne function in **2** is separated from the uridine base by a flexible spacer. Incorporation of building blocks **1** and **2** into 16-mer strands via solid phase DNA synthesis proceeded smoothly albeit with a slight alteration in the phosphoramidite coupling protocol (Table 1). The compatibility of the click reaction was then investigated via the coupling of the ODN series in Table 1 with azides **3-5**. Azides **3-5** were chosen as they represent a small selection of desirable labels. Azido-sugar **3** is a semi-protected aldehyde used for selective Ag-staining;[4a] coumarin azide **4** fluoresces only after triazole formation,[5] and fluorescein azide **5** is a strongly fluorescent molecule used in a variety of biophysical applications.

TABLE 1. ODN series comprising **1** or **2**.

	X = 1
ODN-1	5'-GCG CTG TXC ATT CGC G
ODN-2	5'-GCG CTG XXC ATT CGC G
ODN-3	5'-GCG CXG TXC AXT CGC G
ODN-4	5'-GCG CXX XXX XGT CGC G
	Y = 2
ODN-5	5'-GCG CTG TYC ATT CGC G
ODN-6	5'-GCG CTG YYC ATT CGC G
ODN-7	5'-GCG CYG TYC AYT CGC G
ODN-8	5'-GCG CYY YYY YGT CGC G
ODN-9	5'-TTA ATT GAA TTC GAT TYG GGC CGG AYT TGT TTC
ODN-10	5'-GCA GGC YTCA YGC CAG AAT TAC CAG AAG

The rigid alkyne series ODN-1-ODN-4 was assayed using azides **3-5**. Investigation of the click reaction using the sugar azide **3** and ODN-1 revealed complete conversion (Figure 1a, ODN-1, m/z 5082 [M+Na$^+$]) by MALDI-TOF analysis; however when the alkyne density was increased to two adjacent alkyne click sites (Figure 2a, ODN-2), an additional small amount of the mono-click product (m/z 5110 [M+Na$^+$]) was observed in addition to the fully clicked product (m/z 5297). This result was also consistent with the click reaction between **3** and the ODN containing three alkyne functions (ODN-3) separated by two bases each. A minor amount of the two-clicked adduct (m/z 5320 [M+Na$^+$]) was obtained in addition to the three-clicked adduct (m/z 5508) as the major product. These findings were also supported by the reaction with ODN-4 containing six adjacent alkynes. The major click product then corresponded to only a five-clicked adduct (m/z 5970 [M+2 Na$^+$]) in addition to minor products corresponding to a four (m/z 5784 [M+3Na$^+$]) and six (m/z 6158 [M+Na$^+$]) clicked adducts respectively. Therefore it became apparent that the steric shielding of the alkyne by the DNA backbone (ODN-1-ODN-4) was contributing to the less than optimal labeling yields. In contrast to the ODN-1-ODN-4 series full conversion was observed for ODN-5, ODN-6 and ODN-8 containing the flexible alkyne according to MALDI-TOF analysis (Figure 2b).

Switching to the non-fluorescent coumarin azide **4** provided the added benefit of forming a highly fluorescent click product. Both ODN series (ODN-1-ODN-3 and ODN-5-ODN-8) were assayed against azide **4** and revealed a similar clicking scenario to that observed for azide **3**: the rigid alkyne series (ODN-1-ODN-3) produced less than quantitative conversion to the fluorescent click product, whereas the flexible alkyne series (ODN-5-ODN-8) produced fully labeled products. With the flexible alkyne spacer nucleoside **2**, it is therefore possible to achieve highly reliable and complete functionalisation of ODNs. Gel electrophoresis studies of coumarin-labeled ODN products (ODN-5-ODN-7) are depicted in Figure 3a.

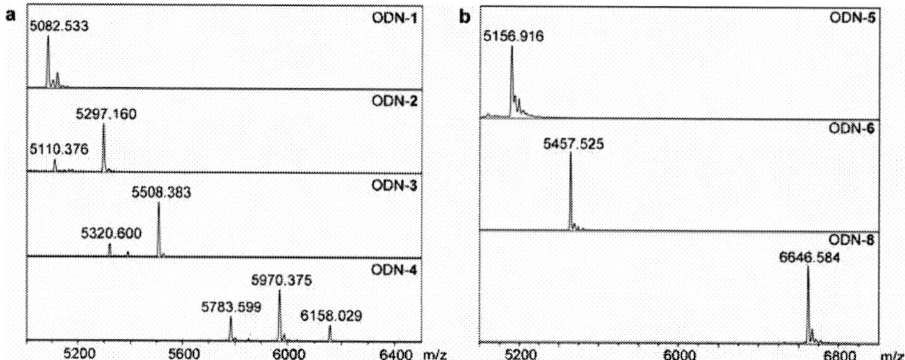

FIGURE 2. MALDI-TOF spectra of **(a)** Click reaction performed with azide **3** and ODNs comprising alkyne **1** with one (ODN-**1**, M.W. 4872), two (ODN-**2**, M.W. 4882), three (ODN-**3**, M.W. 4892) and six (ODN-**4**, M.W. 4902) alkyne functions. **(b)** Click chemistry performed with azide **3** and ODNs comprising alkyne **2** with one (ODN-**5**, M.W. 4952), two (ODN-**6**, M.W. 5042) and six (ODN-**8**, M.W. 5379) alkyne functions.

When click chemistry was performed using the fluorescein azide **5**, the resulting products were easily observed by the difference in electrophoretic mobility as a consequence of a much larger increase in molecular weight (Figures 3b + 3c). As expected, in the case of the ODN-1-ODN-4 series, only partial conversion was observed (Figure 3b). For two click sites (lane 2) and three click sites (lane 3) the product mixtures are clearly visible. The click reaction between **5** and ODN-4 in particular revealed a range of products (Figure 3b, lane 4). Consistent with previous click studies using the ODN-5-ODN-8 series, full-conversion was observed for all strands by gel electrophoresis (Figure 3c). Lanes 1 to 4 show complete conversion of one, two, three and six click sites.

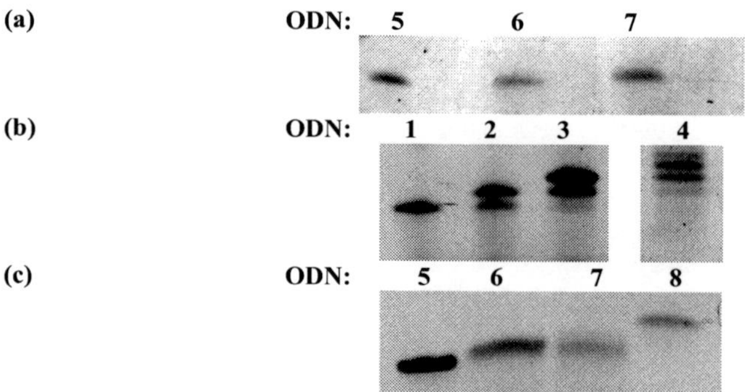

FIGURE 3. Gel electrophoresis of click reactions. **(a)** Click reactions of ODNs containing **2** with coumarin azide **4**; detection using a 460 nm cutoff filter. **(b)** Click reactions of the ODN series containing the nucleoside **1** and the fluorescein azide **5**. **(c)** Click reactions of the ODN series containing the nucleoside **2** and the fluorescein azide **5**

To investigate whether click chemistry can be utilized to label long DNA fragments PCR-primer ODN-9 and ODN-10 containing two click sites were synthesized and subsequently used in PCR to amplify a range of amplicons from two different plasmid templates. Melting point analysis of ODN-9 revealed no destabilization of the DNA strand due to the two modifications.

The resulting amplicons comprise two alkyne moieties at one end of the double strand. The purified DNA was then treated with azide **5** using the CuBr/ligand method.[6] Gel electrophoresis of the reaction showed only a single band corresponding to a fluorescein labeled click product (Figure 4a, lane 2 and Figure 4b lane 2 and 4). An unmodified DNA fragment treated identically showed no fluorescence (Figure 4a, lane 3, Figure 4b lane 1 and 3). No sign of DNA degradation was observed when staining the gels with SYBR Green II or ethidium bromide in either case. Similar results were obtained on PCR fragments up to 2000 base pairs.

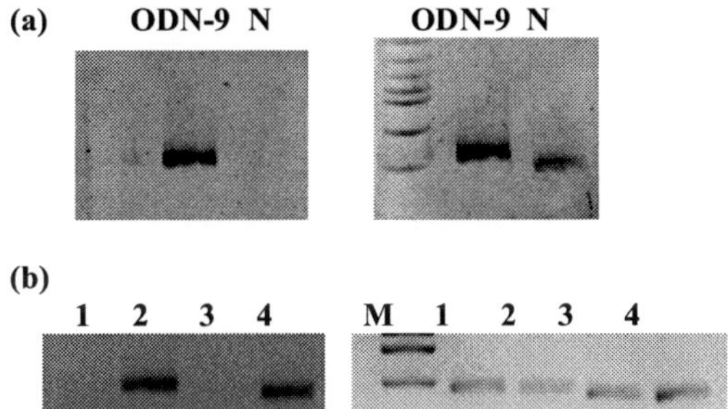

FIGURE 4. Click reactions on the PCR product using fluorescein azide **5**. (a) Click on 300 pb PCR fragment comprising ODN-9 (lane 1: 100bp-marker; lane 2: Alkyne DNA (primer ODN-9); lane 3: control with unmodified DNA; N=natural) The right gel is stained with SYBR Green II; (b) Agarose gels of a Click on a 800 bp PCR product using ODN-10 (lane 2) and a 900 bp PCR product using ODN-9 (lane 4). Lane 1 and 3 are controls using unmodified primers The right gel was run with ethidium bromide.

Using click chemistry to direct metal deposition along DNA templates

In order to probe the scope of directing labeling processes to DNA strands gene-specifically, we investigated whether metal deposition could be directed towards alkyne modified genes up to 2000 base pairs in length. Subsequently, we devised the two-step protocol depicted in Figure 5. In the first step, DNA polymerases are used to introduce acetylene reporter groups into selected genes via the enzymatic incorporation of C5-modified pyrimidine nucleoside triphosphates, such as **6** and **7** (Figure 6).[7] The second step involved reaction of the acetylene reporter groups with aldehyde azides using click chemistry. As a consequence of this derivatization process, the selected genes are now adorned with aldehyde functions. Although the direct incorporation of an aldehyde-modified triphosphate analogue provides a more expeditious route toward aldehyde-modified DNA, their enzymatic incorporation provided mixed results. To evaluate how efficiently both compounds were accepted by various polymerases, we used the *polη* and *polH* genes from yeast and human cDNA as suitable template DNA strands.[8]

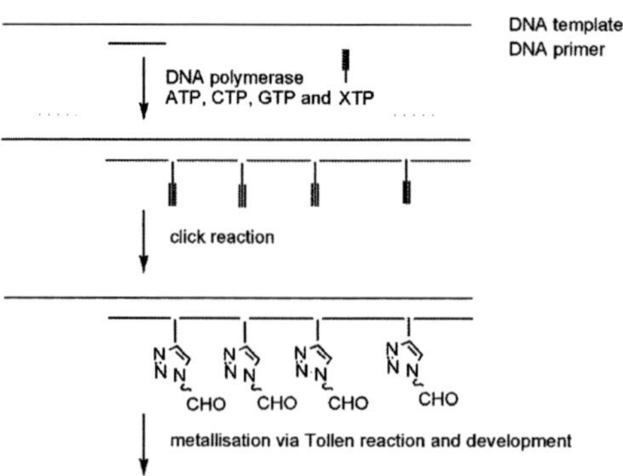

FIGURE 5. Schematic depiction of the selective metallization process.

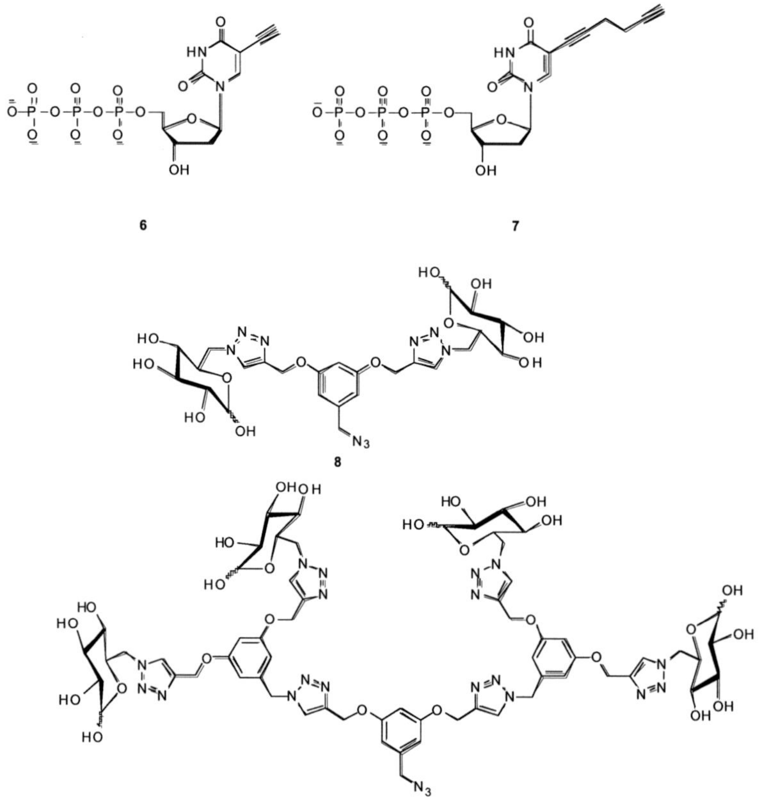

FIGURE 6. Depiction of the molecules **6 - 9** used for this study.

These genes were initially isolated, then cloned into a plasmid, and subsequently used for the PCR.[9] Amplification of both the *polη* and *polH* genes using standard PCR conditions with a mixture of triphosphates (dATP, dCTP, dGTP, and **6/7**) afforded, for both triphosphates (**6** and **7**), full-length amplicons (2142 bp for human *polη* and 318 bp for yeast *polη*) (Figure 7a).[9] In experiments where triphosphate **6** replaced dTTP, full-length amplicons were readily obtained using a variety of commercially available high-fidelity polymerases. For compound **7**, however, full-length amplicons of suitable quantity were only obtained when the *Pwo* polymerase was used. In both examples, the class B *Pwo* polymerase provided the highest yields of PCR products. Enzymatic digestion of the acetylene-decorated *polη* gene amplicon incorporating **7** and subsequent HPLC and mass spectrometric analysis confirmed the complete replacement of thymidine by the nucleoside **2**. In addition, the acetylene-modified DNA strands can also be used as template strands. PCR amplicons generated with **6** or **7** could be used as templates for the PCR using either the four natural triphosphates or a triphosphate mixture containing, again, **6/7** instead of dTTP. Sequencing of the PCR products obtained with the natural set of triphosphates provided the correct base sequence, reflecting the high fidelity of the polymerases when incorporating **6** or **7**.

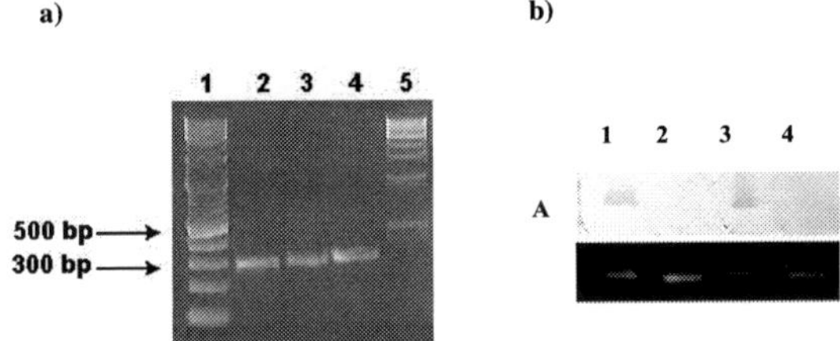

FIGURE 7. PCR assays (*Pwo* pol.) of the *polη* gene from yeast (318 bp). (a) [Lanes 1 and 5: DNA ladder (NEB 2-log; 0.1-10.0 kb). Lane 2: Positive control (using dTTP). Lane 3: PCR using **6**. Lane 4: PCR using **7**. (b) PCR assay incorporating triphosphate **7** [Lane 1: 7.0 ng; Lane 3: 3.5 ng DNA loadings] and dTTP [Lane 2: 7.0 ng, Lane 4: 3.5 ng DNA loadings]. Gel **A** corresponds to treatment with a Tollens solution followed by development, whereas gel **B** corresponds to treatment with the fluorescent stain SYBR Green II

We then investigated whether the Ag deposition process can be directed to just aldehyde-modified DNA. To this end, natural, unmodified DNA (318 nucleobases) and its DNA cognate prepared with either **6/7** were loaded onto a 5% TBE-urea polyacrylamide gel (Figure 7b). The click reaction was then performed directly on the gel by agitation of the DNA-containing gel in a 1:1 MeOH: H2O solution comprising **3**, CuSO₄, and a reducing agent (e.g., TCEP or sodium ascorbate). Washing the gel with an aqueous Ag(NH₃)₂OH solution (Tollens reagent) and subsequent development of the Ag(0) nuclei with a typical developer solution (citric acid and formaldehyde) furnished, only in the case of the aldehyde-modified DNA strands, yellow/brown spots on the gel, indicative of Ag deposition (Figure 7b, Gel A). The whole process of the Tollens reaction and development is performed in just 30 and 3 min, respectively. To prove that the unmodified DNA is indeed present on the gel, we repeated the experiment but now treated the gel with SYBR Green II (Figure 7b, Gel B). This time, both aldehyde-modified and natural DNA bands were observed. The click reaction followed by Ag deposition is highly efficient because the galactose-modified 318-mer DNA was detectable by eye down to 1.3 ng.[4] The click reaction using **1**-modified

DNA is less efficient, in agreement with the data from the small oligonucleotides, as indicated by the lower detection sensitivity of only 3.5 ng. When the aldehyde density on the DNA was significantly increased, by the use of azide containing dendrimers **8** and **9**, the detection limit of **2**-modified DNA was reduced to 0.9 and 0.5 ng, respectively, indicative of an increase in the amount of Ag(0) deposition around the modified DNA.

CONCLUSIONS

In summary, click chemistry is a simple and robust method for the conversion of alkyne-modified DNA into labeled products. Using the alkyne nucleoside **2**, the high density reliably modification of all alkyne sites can be achieved. Additionally, we have developed an efficient and selective method for the deposition of Ag(0) around aldehyde-modified DNA. The modification involves incorporation of acetylene-containing nucleotide triphosphates using DNA polymerases followed by a click reaction that can be efficiently performed directly on a polyacrylamide gel. Using this method, Ag(0) deposition can be confined only to the modified DNA. The ability to insert the acetylene labels enzymatically offers the possibility to exploit the arsenal of molecular biological tools in order to construct conductive DNA nanodevices. Experiments to ascertain whether these metallised DNA constructs are electrically conductive are currently underway.

ACKNOWLEDGMENTS

We thank the Volkswagen Foundation, the Deutsche Forschungsgemeinschaft and BASF AG (Ludwigshafen) for generous financial support. J.G. thanks the Fonds of the Chemical Industry for a pre-doctoral fellowship. G.A.B. thanks the Alexander von Humboldt Foundation for a research fellowship. P. M. E. G. thanks the Cusanuswerk for a doctoral fellowship.

REFERENCES

1. (a) Seeman, N. C. *Nature* **2003**, *421*, 427. (b). Gothelf, K. V, LaBean, T. H.*Org. & Biomol. Chem.* **2005**, *3*, 4023. (c) Rosi, N.L,. Mirkin, C. A *Chem. Rev.* **2005**, *105*, 1547. (d) Liu, B., Bazan, G. C. *Chem. Mat.* **2004**, *16*, 4467. (e) . Adleman, L. M *Science* **1994**, *266*, 1021. (f). Winfree E *et al.*, *Nature* **1998**, *394*, 539. (g) Seeman, N. C.*Trends. Biochem. Sci.* **2005**, *30*, 119. (h). Yan H *et al.*, *Science* **2003**, *301*, 1882.
2. (a) Braun, E.; Eichen, Y.; Sivan, U.; Ben-Yoseph, G. *Nature* **1998**, *391*, 775-778. (b) Keren, K.; Krueger, M.; Gilad, R.; Ben-Yoseph, G.; Sivan, U.; Braun, E. *Science* **2002**, *297*, 72-75. (c) Keren, K.; Berman, R. S.; Buchstab, E.; Sivan, U.; Braun, E. *Science* **2003**, *302*, 1380-1382. (d) Deng, Z. X.; Mao, C. D. *Angew. Chem.-Int. Ed.* **2004**, *43*, 4068-4070. (e) Richter, J.; Seidel, R.; Kirsch, R.; Mertig, M.; Pompe, W.; Plaschke, J.; Schackert, H. K. *Adv. Mat.* **2000**, *12*, 507. (f) Mertig, M.; Ciacchi, L. C.; Seidel, R.; Pompe, W.; De Vita, A. *Nano Lett.* **2002**, *2*, 841-844. (g) Monson, C. F.; Woolley, A. T. *Nano Lett.* **2003**, *3*, 359-363.

3. (a) Kolb, H. C.; Finn, M. G.; Sharpless, K. B. *Angew. Chem.-Int. Ed.* **2001**, *40*, 2004-2021. (b) Rostovtsev, V. V.; Green, L. G.; Fokin, V. V.; Sharpless, K. B. *Angew. Chem.-Int. Ed.* **2002**, *41*, 2596-2598.

4. (a) Burley, G. A.; Gierlich, J.; Mofid, M. R.; Nir, H.; Tal, S.; Eichen, Y.; Carell, T. *J. Am. Chem. Soc.* **2006**, *128*, 1398-1399. (b) Gierlich, J.; Burley, G. A.; Gramlicj. P.; Hammond, D.M. ; Carell, T. *Org. Lett.*, **2006**, in press.

5. Sivakumar, K.; Xie, F.; Cash, B. M.; Long, S.; Barnhill, H. N.; Wang, Q. *Org. Lett.* **2004**, *6*, 4603-4606.

6. (a) Beatty, K. E.; Xie, F.; Wang, Q.; Tirrell, D. A. *J. Am. Chem. Soc.* **2005**, *127*, 14150-14151. (b) Link, A. J.; Vink, M. K. S.; Tirrell, D. A. *J. Am. Chem. Soc.* **2004**, *126*, 10598-10602. (c) Link, A. J.; Tirrell, D. A. *J. Am. Chem. Soc.* **2003**, *125*, 11164-11165.

7. (a) Sakthivel, K.; Barbas, C. F. *Angew. Chem.-Int. Ed.* **1998**, *37*, 2872-2875. (b) Held, H. A.; Benner, S. A. *Nucleic Acids Res.* **2002**, *30*, 3857-3869.

8. (a) Johnson, R. E.; Prakash, S.; Prakash, L. *Science* **1999**, *283*, 1001-1004. (b) Hubscher, U.; Maga, G.; Spadari, S. *Ann. Rev. Biochem.* **2002**, *71*, 133-163.

9. For examples of click chemistry performed on DNA, please consult the following references: (a) Devaraj, N. K.; Miller, G. P.; Ebina, W.; Kakaradov, B.; Collman, J. P.; Kool, E. T.; Chidsey, C. E. D. *J. . Am. Chem. Soc.* **2005**, *127*, 8600-8601. (b) Weller, R. L.; Rajski, S. R. *Org. Lett.* **2005**, *7*, 2141-2144.

Specific Metal Deposition onto Immobilized Metal Nanoparticles Studied on Single Particle Level

Grit Festag, Thomas Schüler, Andrea Steinbrück, Andrea Csáki, Wolfgang Fritzsche

Institute for Physical High Technology, Photonic Chip Systems Department, Jena, Germany

Abstract. Specific metal deposition on metal nanoparticles plays an important role in molecular construction as well as for microscopic visualization and signal enhancement in various bioanalytical methods. To characterize this process we studied surface-immobilized gold nanoparticle ensembles at single particle level by atomic force microscopy monitoring the very same particle arrangements after each enhancement step. Thereby, enhancement solutions of different composition are studied regarding their enhancement rate as measured by the growing height of the particles at single particle level. Other features of interest were the influence of the diameter of the seed particle on the growth process and the influence of exchange of enhancement solution. For the applied conditions (low nanoparticle surface concentration subjected to fresh enhancement solution 2 min each) we observed a linear growth regime, which was dependent on the particles' seed diameter and differed considerably in enhancement rate, homogeneity, and specificity among the examined enhancement solutions. We extended these studies to an investigation of enzymatic metal deposition where an enzyme complex (horse radish peroxidase) catalyzes the growth of a metal layer.

Keywords: AFM, gold nanoparticles, metal enhancement, metal deposition, enzyme
PACS: 61.46.Df; 81.16.-c; 81.16.Be

INTRODUCTION

Beside their use as labels in microscopy and bioanalytics, metal nanoparticles are extremely effective at enhancing the scattering signal from Raman-active molecules [1]. In addition, they can be further used as building blocks in the formation of nanostructured material. The highly specific deposition of a metal shell onto these particles (based on autocatalytic reduction of metal salts at the particle surface [2]) represents a basic approach in DNA-based molecular construction as well as in nanoparticle-based labeling methods. To synthesize designer particles with defined optical properties and to control this metal deposition process one should know about the characteristics of nanoparticle growth as well as the features of different metal deposition solutions. Although it is known that increasing enhancement time and higher temperatures lead to an increased particle growth, a detailed understanding is still lacking.

In order to characterize the underlying growth process we studied the influence of the kind of enhancement solution, incubation time, and seed size by monitoring the heights of the very same particle ensembles during a stepwise metal deposition with atomic force microscopy (AFM). Since there are some drawback with nanoparticle-based labeling techniques, for example metal deposition also at electrode structures, we also investigated an enzymatic approach for metal deposition with regard to the increase of particle growth depending on different enhancement times by atomic force microscopy.

MATERIALS AND METHODS

Particle Samples

Different-sized gold nanoparticles (15, 30, and 60nm in diameter, subsequently named as Au15, Au30, and Au60, respectively) were nonspecifically immobilized on silicon oxide chips by incubating with the colloidal solutions over night at room temperature (final concentration $2x10^{11}$ particles per ml according to manufacture's instructions; BBinternational, Cardiff, UK), leading to well-separated particles on the chip surface. The resulting particle density on the surface varied from of 0.7 (Au60) to 4 (Au15) particles/μm^2 leading to well-separated particles.

The method of enzymatic metal deposition requires horse radish peroxidase (HRP) as catalyst, and a three component enhancement kit (EnzMet, ultrasensitiv, Nanoprobes, Yaphank, NY [3]). We worked with streptavidin conjugated HRP from Sigma (1,1mg/ml ultrasensitive, Sigma-Aldrich GmbH, Taufkirchen, Deutschland), commonly used in microarray technology as marker for target biomolecules in a 1:200 dilution in double destilled water.

Enhancement Procedure

The metal enhancement of the immobilized particles was achieved by a 2min stepwise incubation (up to 10min enhancement time in total) with different metal salt/reducing agent solutions at room temperature in darkness using different commercial as well as homemade silver and gold enhancement kits (Sigma-Aldrich, BBInternational, Aurion; silver acetate/hydroquinone [4], silver nitrate/hydroquinone [5], and tetrachloroaurate/hydroxylamine [6], subsequently abbreviated as Sigma, BBI, Aurion, AgAc, $AgNO_3$, and $HAuCl_4$, respectively). To reveal a possible limitation of the metal salt/reducing agent, the samples were additionally enhanced in longer incubation steps, twice for 5min. The two components of the enhancement kits were mixed directly before applying them as a droplet onto the substrate with the bound particles.

The diluted enzyme polymer was unspecifically bound to a mica surface by 5min incubation at room temperature, and was subsequently dried under a nitrogen stream. To avoid the loss of enzymatic activity due to longer dryness, the first component (so called solution A) of the EnzMet enhancement kit was immediately applied to the surface. After another 5min incubation time a mixture of the other two components (solution B &C) was added. The reaction was stopped after 1, 2, 3, 4 or 5min enhancement with water. Because there is no possibility to restart the metal deposition procedure after drying, each enhancement step is represented by a different substrate.

AFM Characterization and Image Analysis

To follow up the enhancement process of the gold nanoparticles with nanometer resolution at single particle level we monitored the very same particle arrangements by AFM (Dimension 3100, Digital Instruments, Santa Barbara, CA; NanoScope, Version 5.12) before and after each enhancement step. The mica substrate with the immobilized HRP was analyzed with a Multimode AFM (Digital Instruments, Santa Barbara, CA, NanoScope, Version 4.31ce). To monitor the metal deposition and follow the growth process we analyzed the particle heights encoded by the brightness in the top view AFM images with the open-source image processing and analysis programs Image J and Image SXM using scan-specific parameters extracted from the original image files.

RESULTS AND DISCUSSION

Stepwise metal enhancement of immobilized particles and enzymes

AFM imaging enables the investigation of the specific metal deposition at single particle level with nanometer resolution (especially regarding height). To follow up the stepwise growth process of individual gold nanoparticles, several areas on the chip were chosen and ensembles with the very same particles were visualized before and after each enhancement step. Fig. 1 shows plain 30nm-sized gold nanoparticles and their growth after 4 and 8min silver and gold enhancement as examples. The section analyses of a selected particle reveals the increase in height due to the metal deposition onto the particles and indicate the differences in growth intensity among the different enhancement solutions.

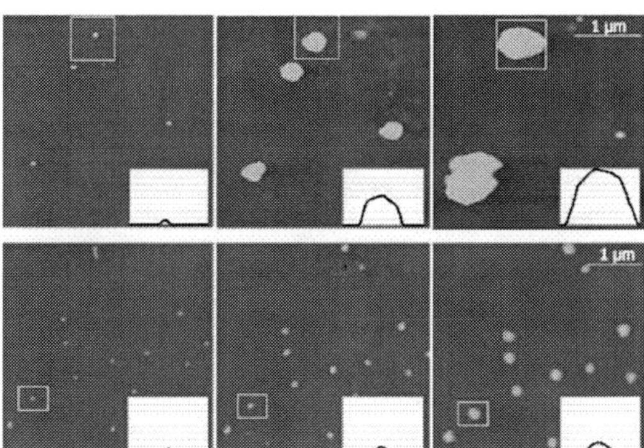

FIGURE 1. Stepwise metal deposition on gold nanoparticles (Au30) with reductive silver (top) and gold enhancement (bottom). The AFM images show the very same particles before (left), after 4min (center), and 8min (right) metal enhancement. The insets show cross sections of a selected particle (600nm in width, 400nm in height, respectively).

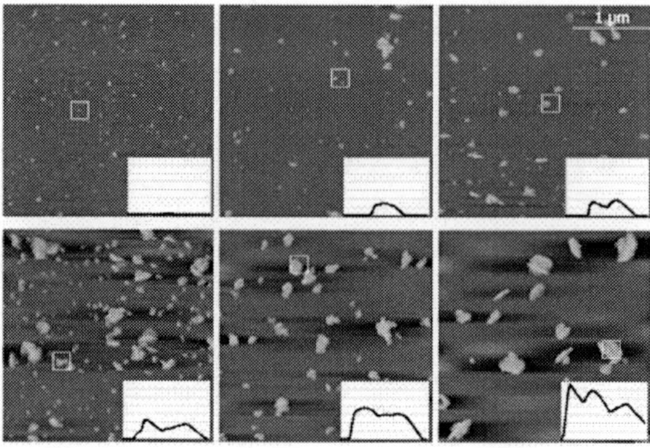

FIGURE 2. Stepwise enzymatic metal deposition starts with the untreated enzyme polymer (top left) followed by deposited particles in 1min steps up to 5min (bottom right). The cross sections of the selected particles in every image demonstrates both the increase of the particle height and the perimeter at different enhancement time.

Particle Height/Growth Rate

Nanoparticle enhancement

To characterize the growth characteristics by specific metal deposition onto the nanoparticles, the measured mean particle height was examined in the course of the enhancement process (Fig. 3). In general, the particles showed a linear growth behavior, which pointed to a constant shell growth without any limitation effects. This is understandable considering the applied experimental conditions, such as low particle surface concentration and a quasi continuous solution exchange.

A comparison of the applied commercial and homemade metal enhancement solutions revealed differences in their enhancement effect. While most of the different solutions showed similar enhancement intensity (final particle heights around 77nm±11nm), the BBI silver enhancement kit resulted in a significant higher particle size (437nm after 10min enhancement) (Fig. 3 left; note the different height scales). This corresponded to a 11-fold higher growth rate.

Furthermore, the enhancement of different-sized gold nanoparticles (Au15, Au30, and Au60) revealed a seed size-dependent particle growth: the larger the seed particles the faster the growth (Fig. 3 right).

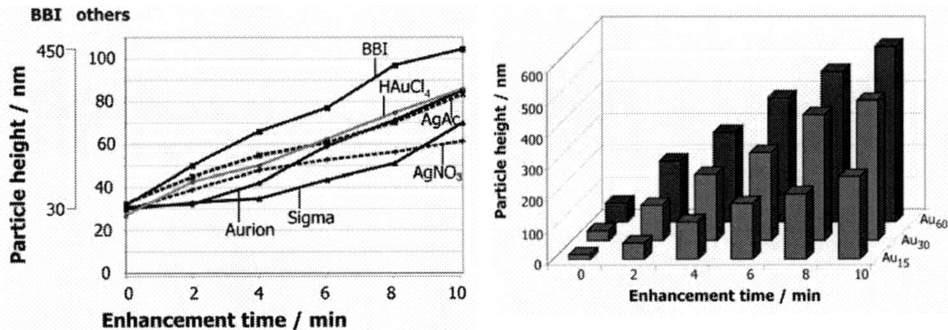

FIGURE 3. The growth of gold nanoparticles dependent on the applied enhancement solution (Au30, left) and particle seed diameter (Au15, 30, Au60; right). The particles grew linearly when enhancement solutions were freshly added each 2min. The final particle height was strongly dependent on the applied enhancement solution (left) and the particle seed size (right): the larger the seed particles the faster the growth.

To reveal possible limitation effects regarding the enhancement components supply, we extended the incubation/enhancement intervals (Fig. 4).

The linear growth by constantly adding freshly mixed solutions to the chip was decelerated when enhancement intervals were prolonged from 2 to 5min, which pointed to a beginning limitation of metal salts and/or reducing agents (Fig. 4 left).

This assumption was strongly supported by comparing the multiple-step approach versus a one-step enhancement procedure without any solution exchange (Fig. 4 right). The lack of solution exchange led to a significant lower final particle height compared to the quasi-continuous addition of enhancement solutions (210±23 vs. 600±76nm) suggesting that the reagent were consumed during the enhancement process.

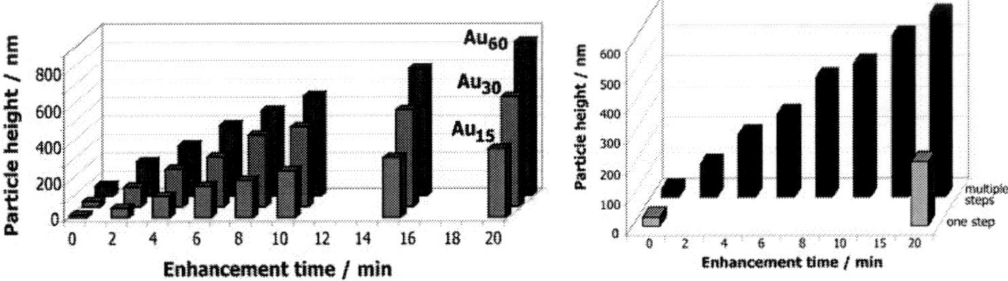

FIGURE 4. The linear growth was decelerated by additional enhancement in 5min-steps pointing to a beginning diffusion limitation (left). Enhancement without freshly added solutions (one-step) was strongly inhibited compared to the multiple-step approach because metal salts and/or reducing agent were consumed during the enhancement process (right).

To study the enzyme-based particle growth for various enhancement times we measured particles at every enhancement step and characterized them by height measurements. The mean values show a linear growth (Fig.3 right) but there is a wide range of particle sizes for the different deposition steps (Fig.3 left). However, the maximum height grows with increasing enhancement time. We assume that these differences are caused by different sizes among the untreated enzyme complex.

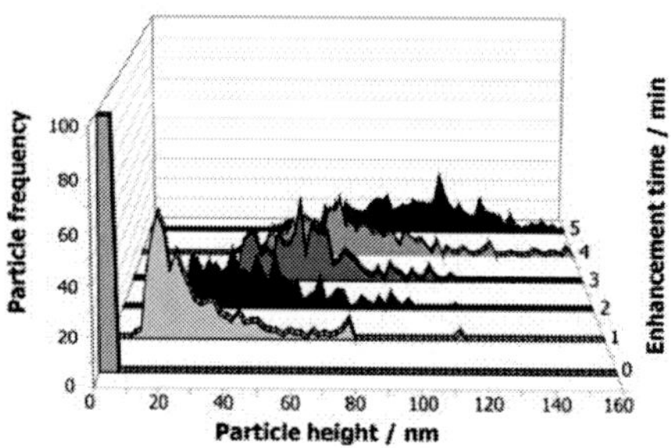
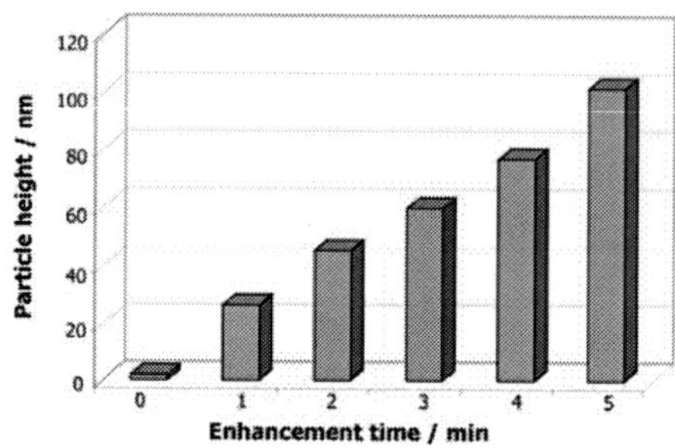

FIGURE 5. The diagram shows the particle height distribution (left) at different enhancement times for the enzyme complex (horse radish peroxidase). The deposition of silver was measured every minute as height. A linear growth becomes apparent (right).

Enhancement Specificity

Beside their different enhancement efficiency, the applied enhancement solutions targeting nanoparticles differed considerably in their specificity (Fig. 6). One the one hand, the measured AFM images revealed a very specific enhancement activity of only the metal nanoparticles leaving the background free of additional metal deposition (BBI silver and $HAuCl_4$ gold enhancement). On the other hand, we monitored heavy non-specific background signals (Sigma and $AgNO_3$ silver enhancement).

Number of Particles/Particle Loss

Finally, by monitoring the very same particle arrangements over the enhancement process, we observed an increasing loss of particles, i.e. a displacement from the chip surface (Fig. 7), which is understandable for the non-specifically immobilized particles. But interestingly, the differences in the relative number of available particles varied notably among the applied enhancement solutions, which could be due to different ionic strength of the solutions. Additionally, the particle displacement could be caused by mechanical forces during the repeating washing/drying processes. Thereby, larger particles should be more sensitive to shear forces, which we observed for the different-sized nanoparticles, where the particle loss was elevated with an increased particle seed size (Fig. 7 left). The influence of mechanical forces by repeating chip handling was strongly supported by comparing the one-step with the multiple-step enhancement approach (Fig. 7 right). We found a large difference in the relative particle number after 20min one-step and stepwise enhancement (98% vs. 27%, respectively).

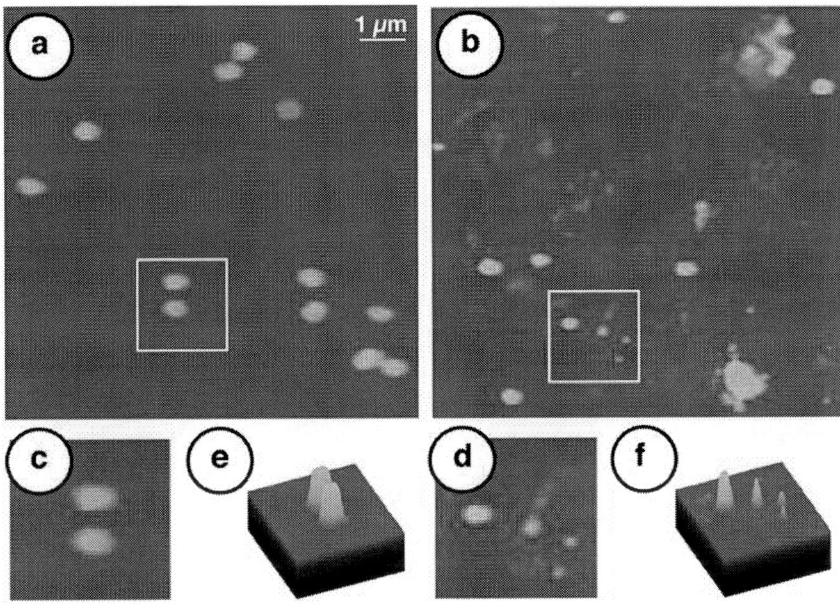

FIGURE 6. Enhancement of gold nanoparticles (Au30) by gold (left, HAuCl$_4$) and silver (right, AgNO$_3$) enhancement for 5x2min. The AFM images show an overview (a,b), a detailed look at single particles (c,d), and a surface plot of these particles (e,f).

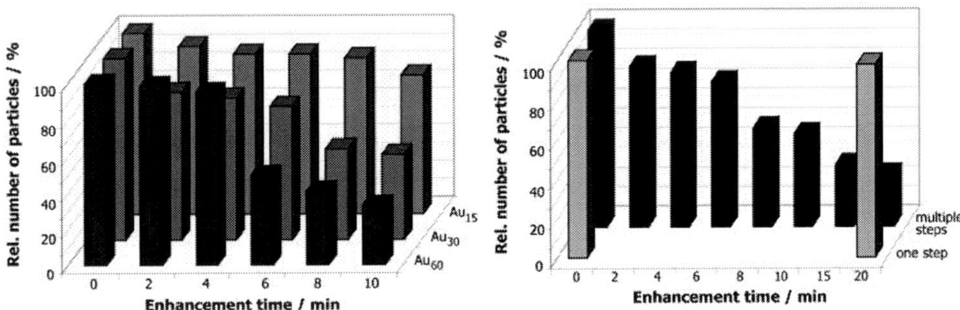

FIGURE 7. In the course of the enhancement process a loss of particles was found due to repeated incubation/drying processes, which seemed to be dependant on the particle seed size (left). The comparison of the one-step vs. multiple-step approach supported the assumed influence of mechanical/shear forces on particle displacement from the surface (right).

CONCLUSION/ OUTLOOK

The processes of autocatalytic as well as enzymatic metal enhancement could be characterized using precise measurements of particle heights by single-molecule AFM imaging as base for a statistical study.

For the utilized systems, the growth rate was quite comparable in the range of tens of nanometers per minute. The low variations in the case of autocatalytic enhancement are probably due to their spherical geometry resulting in an anisotropic growth, in contrast to larger variations in the height distribution for the enzymatic process, that could be caused by an anisotropic growth based on the flake-like substructure and/or the presence of enzyme multimers (beside monomers) in the studied ensemble. Further studies will include the volume of the complexes in order to reach a better description of the growth kinetics, and the investigation of combined enhancement processes.

ACKNOWLEDGMENTS

We thank J. Hull for the Nanoscope™ files import plug-in of the image processing and analysis software Image J and Nanoprobes. This work was supported by the DFG (FR 1348/5-2) and the European Project NUCAN (NMP STREP 013775).

REFERENCES

1. D. A. Stuart, A. J. Haes, C. R. Yonzon, E. M. Hicks, and R. P. Van Duyne, *IEE Proc.-Nanobiotechnol.* **152** (1), 13 (2005).
2. G. W. Hacker, L. Grimelius, G. Danscher, G. Bernatzky, W. Muss, H. Adam, and J. Thurner, *J Histotechnol* **11**, 213 (1988).
3. R. Möller, R. D. Powell, J. F. Hainfeld, and W. Fritzsche, *Nano Lett* **5** (7), 1475 (2005).
4. G.-J. Zhang, R. Möller, A. Csaki, and W. Fritzsche, presented at the DNA based molecular construction, Jena, Germany, 2002 (unpublished).
5. P. M. Heidrich, in *Laborjournal* (2004), Vol. 03, pp. 62.
6. Y. Weizmann, F. Patolsky, P. I., and I. Willner, *Nano Letters* **4** (5), 787 (2004).

ELECTRICAL FIELD

AND CONDUCTIVITY

Quantifying DNA Dielectrophoresis

My-Linh Du, Frank F. Bier and Ralph Hölzel

Fraunhofer Institute for Biomedical Engineering, Molecular Bioanalytics & Bioelectronics, Arthur-Scheunert-Allee 114-116, Bergholz-Rehbrücke, 14558 Nuthetal, Germany

Abstract. A system has been developed for the quantitative determination of the dielectrophoretic response of DNA in solution. Local concentration of fluorescently labeled DNA has been measured by microscopical observation and subsequent evaluation of the acquired CCD images. The influence of frequency, amplitude and modulation of the applied field onto the dielectrophoretic collection of linearized pBlueScript dsDNA has been investigated between 10 kHz and 20 MHz.

Keywords: Dielectrophoresis, DNA, electrode.
PACS: 89.20.-a, 87.80.-y, 87.14.Gg, 87.50.Rr

INTRODUCTION

DNA is increasingly employed for the construction of nanometre sized objects [1, 2]. Although mainly self-assembly processes are exploited distinct spatial manipulation is desirable, e.g. for connecting different parts of the construction to different electrodes. For this purpose dielectrophoresis (DEP) can be used, in which inhomogeneous electric fields attract polarizable particles and molecules towards regions of highest field strength, usually towards electrode edges [3-7]. It has successfully been used for the separation of carbon nanotubes [8, 9]. In order to develop this method for molecular manipulation to a similar extent as it is employed now for lab-on-a-chip systems for biological cells [7, 10], a better understanding of the underlying phenomena is needed. For this purpose an experimental setup has been developed that allows a quantitative investigation of the dielectrophoretic response of macromolecules.

MATERIALS & METHODS

Electrode Chamber

Interdigitated electrodes have been prepared from commercially available surface acoustic wave (SAW) resonators [11]. Here the type R2633 (Siemens/Matsushita) has been used, which consists of a quartz substrate being 4 mm long, 1 mm wide and 0.5 mm high. Its metal case served as a shield against contaminations and was opened

CP 859, *DNA-Based Nanoscale Integration: International Symposium,* edited by W. Fritzsche
© 2006 American Institute of Physics 978-0-7354-0357-4/06/$23.00

by machining before use. Electrical connections were made by placing the resonator pins into integrated circuit sockets soldered onto a printed circuit board. This setup guaranteed secure electrical contacts and, at the same time, allowed a simple change of electrodes. The resonator contains two electrode pairs each consisting of 2 x 35 interdigitated aluminium strips of 800 μm length, 2.3 μm width and 0.3 μm height leaving rectangular gaps of 1.7 μm width between the electrodes.

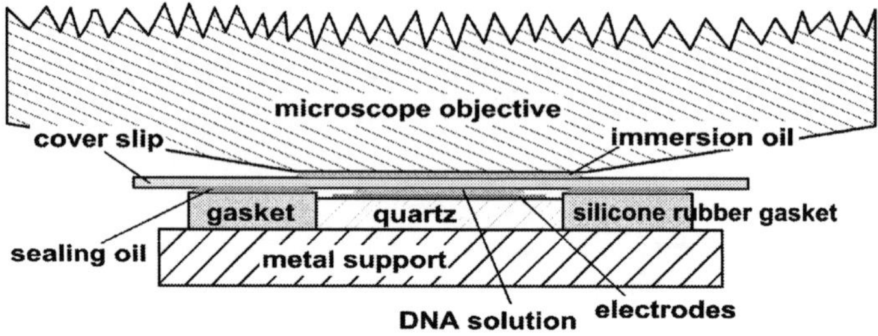

FIGURE 1. Dielectrophoresis chamber. 2 μl of DNA solution are pipetted onto interdigitated electrodes on a quartz substrate. Evaporation is prevented by a silicone rubber gasket, sealing oil and a cover slip.

In order to confine a defined fluid volume a silicone rubber sheet of 0.5 mm thickness was properly cut to serve as a sealing gasket (Fig. 1). Cutting was done either by hand with a scalpel or by a CO_2 laser plotter (Epilog Laser Legend 24TT). The gasket was mounted onto the resonator's bottom with cyanoacrylate glue. A 2 μl drop of the test solution was pipetted onto the electrodes. Sealing was done with a cover slip and by placing oil between cover slip and gasket. This setup allows the use of microscope objectives with a short working distance and prevents evaporation of the small test volume over several hours. Since the chamber volume exceeds the sample volume, the DNA solution only comes into contact with the electrodes, the quartz support and the cover slip. These materials can be thoroughly cleaned thereby helping to keep the DNA solution's electrical conductivity as low as possible despite its small volume.

When using high resolution objectives immersion oil has to be placed between cover slip and objective. This introduces mechanical coupling that can lead to disturbing fluid streaming when focussing or moving the xy-stage. Therefore a sealing oil was chosen whose viscosity substantially exceeds that of the immersion oil but still allows proper removal of the cover slip after the experiment.

Electronic Setup

Sinusoidal signals in a frequency range between 10 kHz and 20 MHz were delivered by a function generator (Wavetek Model 193, Fig. 2). Frequencies were determined by a counter (Voltcraft MXG-9810). Voltages were available up to 8 V_{RMS} (RMS: root-mean-square). They were determined with a demodulator probe and a DC voltmeter (Keithley DMM 191), which had been calibrated by a 1 GHz oscilloscope (Agilent MSO6104A). For a qualitative control of the applied signal another oscilloscope (Hameg HM307) was connected in parallel to the demodulator throughout the experiment. The function generator's built-in auxiliary generator served to modulate the main output, i.e. to switch the signal on and off at a rate of 100 Hz, with variable duty cycle. During the experiment duty cycle was monitored with the voltmeter having been calibrated for this purpose again by the 1 GHz oscilloscope. To allow proper adjustment of the exciting signal without affecting the sample, electrodes were connected to the signal source by a manual switch that was closed only when needed. All connections were made with coaxial cables.

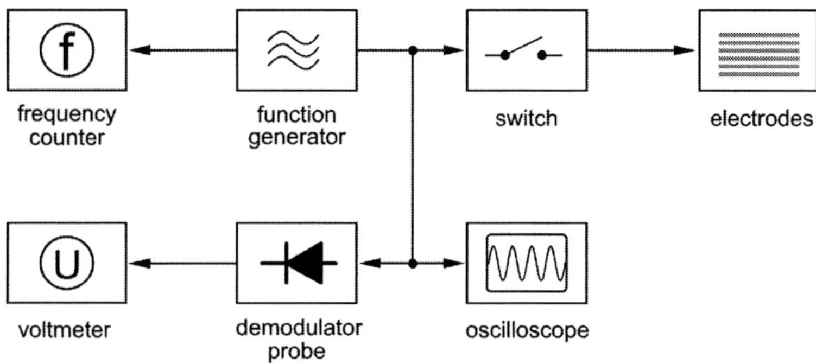

FIGURE 2. Electronic setup. The AC signal is monitored by an oscilloscope and by a voltmeter through a demodulator. A manual switch permits excitation of the electrodes only when needed.

Observation And Data Acquisition

DNA collection was observed under a fluorescence microscope (Olympus BH-2) using a x100 objective (NA 1.25) and a photo ocular (NFK 2.5xL). Images were acquired by a Peltier cooled colour CCD camera (DVC 1312C-LV-TE) that was computer controlled by the software C-View (DVC). The friction adjustment of the microscope's focus knob was tightened to guarantee a constant focus during an experiment.

To determine the dielectrophoretic response an image was acquired before each field application, then the field was switched on, and after 1 min a second image was taken. Then the field was switched off, followed by a break of 3 min to allow even distribution of DNA by diffusion. This period also was used to change the field parameters. Then the acquisition cycle started again. Exposure times were 5 s. To minimize bleaching the sample was only illuminated during image acquisition.

Sample Preparation

As a well defined DNA the vector pBlueScript (Stratagene) was used. It is a double stranded circular DNA consisting of 2961 base pairs corresponding to a physical length of 1.0 μm. It was prepared from a bacterial cell culture and linearized with the restriction enzyme Eco RI. The sample was purified using an Invisorb Spin PCRapid Kit (Invitek) leading to a concentration of about 20 μg/ml. DNA was fluorescently labeled with the intercalating dye Picogreen (Molecular Probes) by mixing identical volumes of DNA solution and 1:200 diluted dye stock solution. To keep the electrical conductivity of the solution as low as possible it was dialysed three times against ultra pure water (Slide-A-Lyzer Mini dialysis kit, Pierce). 2 µl of the dialysed solution were used in each experiment.

Determination Of Sample Conductivity

After completion of an experiment the sample was transfered to a micro conductivity cell prepared from interdigitated gold electrodes [12]. Series resistance was measured at 10 kHz and 0.5 V using an impedance meter (Hioki 3532-50) and was converted to conductivity by applying a calibration performed beforehand by solutions of known conductivity. Despite its small volume the effort was undertaken to measure the actual sample because it should better represent the situation during the experiment than measuring the stock solution. This is because tiny impurities can lead to a significant increase in conductivity due to the small test volume. Above that, since conductivity can only increase in the course of the experimental procedure, e.g. by impurities, evaporation and CO_2 dissociation, measuring it after the DEP experiment will give an upper conductivity limit whilst measuring it before or in the stock solution can only deliver a lower limit value.

Determination Of DNA Response

As a measure of the dielectrophoretic response of DNA the increase in fluorescence during electric field application was taken. For this the fluorescence images were taken and a region of interest was chosen. For each image the corresponding average intensity value was determined with the image processing programme Astra Image (Phase Space Technology). The intensity difference between the images before field application and after field application was then taken as the relevant parameter.

RESULTS & DISCUSSIONS

In order to test the basic assumption that the increase in fluorescence intensity is a reliable measure of the dielectrophoretic response of macromolecules in solution, experiments were performed at constant frequency and amplitude of 1 MHz and 11 V_{pp}, resp. However, the generator's output was modulated, i.e. switched on and off repeatedly, at a rate of 100 Hz and the on-time (duty cycle) was varied from 7% to 95%. For a true representation of DEP by fluorescence changes a linear relationship is to be expected. The results are shown in Fig. 3. Up to a duty cycle of about 50% the fluorescence response was linear with a regression coefficient of $r^2 = 0.994$. For longer field-on times the response exhibited saturation. Background subtraction did not lead to significant changes. Saturation partially is a consequence of the camera being overexposed at some pixels in the region of interest. At the same time the vicinity of the electrodes becomes depleted from DNA limiting the supply of further molecules. The background level was found to increase with duty cycle. This is a consequence of the limited period of 3 min between field applications, which seems to be somewhat too short to allow complete distribution of DNA by diffusion after being concentrated at the electrodes.

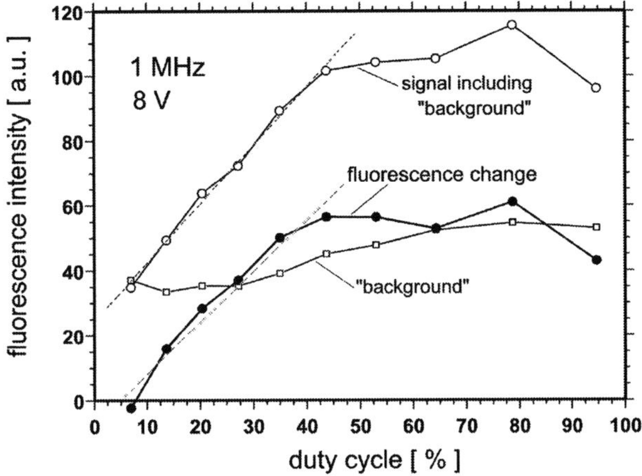

FIGURE 3. Influence of the DEP field's duty cycle on the fluorescence signal. Open circles: absolute fluorescence intensity, open squares: fluorescence before field application, closed circles: fluorescence change during field application. Broken lines: linear regression for duty cycles lower than 50%. Medium conductivity 250 μS/cm.

The intensity profile across the electrode gap (Fig. 4) shows that maximal fluorescence is found close to the electrode edges. However, it is not exactly in the gap between the electrodes rather than on the electrode surface. This might be a consequence of AC electro-osmosis [13] leading to additional fluid streaming from the gap towards the electrode centre.

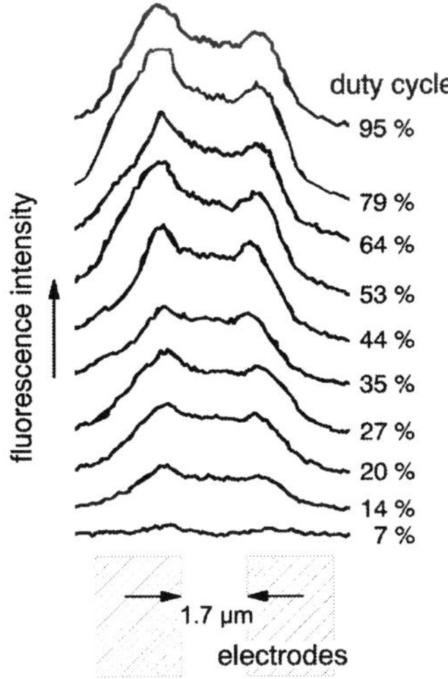

FIGURE 4. Intensity profile across the electrodes for various duty cycles. Fluorescence maxima are found on top of the electrodes close to the edges.

Other work on dielectrophoresis of DNA showed maximum DNA concentration exactly in the gap [12, 14, 15]. In these studies the length of the DNA usually exceeded the gap width allowing the molecules to be attracted at both ends bridging the gap. In this case fluid streaming might even push DNA molecules down to the glass surface. I this study the DNA's length of 1.0 µm is much shorter than the gap width of 1.7 µm. A similar effect has been reported for sub-micrometre latex spheres [16], which have been concentrated by DEP on top of electrodes close to their edges.

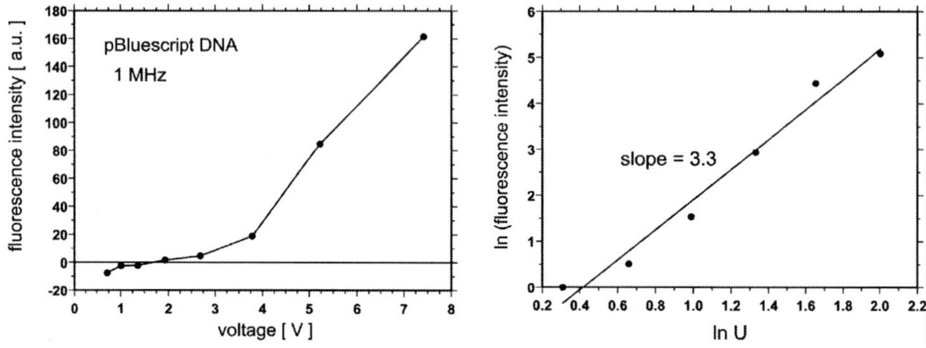

FIGURE 5. Dependence of dielectrophoretic response of pBlueScript DNA (2960 bp) on electrode voltage. Left: Linear representation, right: double logarithmic plot revealing a slope of 3.3. Medium conductivity 250 µS/cm.

The influence of field strength on DNA collection was investigated by varying the electrode voltage from 0.7 to 7.4 V_{RMS} at 1 MHz (Fig. 5). As expected an increase in field strength led to stronger fluorescence. However, at very low voltages fluorescence decreased during field application. This could be a consequence of bleaching the fluorophores, which is not obvious at higher DNA concentration. Representing the data in a double logarithmic plot (Fig. 5) shows a good linearity between the logarithmic values. However, the slope amounts to 3.3 meaning that the DEP response increases with at least the cube of the applied voltage. This is in contradiction to the generally accepted model of induced dipoles being responsible for dielectrophoresis [3, 6, 17]. Perhaps this is a consequence of multipoles having to be taken into account [18] or of additional phenomena like AC electro-osmosis [13] and electrothermal fluid flow [19], which could be dependent on the applied signal with other relations than a square law.

When performing dielectric investigations on microscopical particles like biological cells usually most information can be deduced from spectra. Such a variation of the field frequency has been done here on pBlueScript DNA between 10 kHz and 20 MHz (Fig. 6). There is a clear maximum between 100 kHz and 3 MHz coinciding well with the literature, where frequencies around 1 MHz have been used for the dielectrophoretic manipulation of DNA [5, 14, 20]. Dielectrophoretic elongation studies of λ-DNA [21] also show a broad maximum between 100 kHz and 1 MHz. Above 10 MHz negative dielectrophoresis occurs. The observed decrease in DEP response with increasing frequency from a maximum value (here at 300 kHz) is similar to that found for microscopical particles, which is well understood [6, 17]. These theories predict a constant value of dielectrophoretic forces at low frequencies decreasing to a lower, also constant value at higher frequencies.

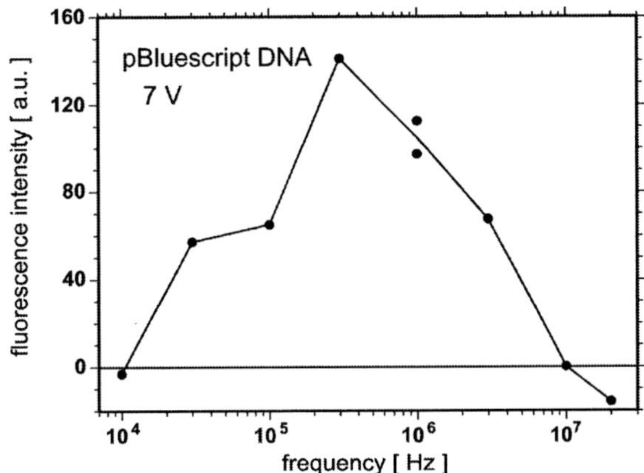

FIGURE 6. Dielectrophoretic spectrum of pBlueScript DNA at 7 V$_{RMS}$ and a medium conductivity of 50 µS/cm.

The reduced response below 100 kHz found here is not covered by these models. However, a similar observation has been made in the dielectrophoresis of polystyrene microparticles and has been explained by the electric double layer covering the electrodes [22]. Its action is analogous to that of a capacitor blocking low frequency signals, i.e. it effectively reduces the electric field in the bulk at low frequencies.

CONCLUSIONS

A system has been developed for the quantitative investigation of the dielectrophoretic response of macromolecules. It will allow a better understanding of the parameters influencing molecular DEP and, hence, will open the way to more complex molecular manipulations by AC electric fields like sorting of molecules and construction of supramolecular assemblies. The apparent and unexpected cubic voltage dependence needs further investigation. Measuring ionic strength in the actual sample volume should be implemented rather easily by using the second electrode pair or the DEP electrodes themselves.

ACKNOWLEDGMENTS

We thank Mandy Lorenz and Markus von Nickisch-Rosenegk for help with the pBlueScript preparation. Support by the European Commission (6th Framework Programme, STRP 013775) is thankfully acknowledged.

REFERENCES

1. P. W. K. Rothemund, *Nature* **440**, 297-302 (2006).
2. N. Seeman, *Trends Biochem. Sci.* **30**, 119-125 (2005).
3. H. A. Pohl, *Dielectrophoresis*, Cambridge: Cambridge University Press, 1978.
4. C. F. Gonzalez and V. T. Remcho, *J. Chromatography A* **1079**, 59-68 (2005).
5. R. Hölzel and F. F. Bier, *IEE Proc.-Nanobiotechnol* **150**, 47-53 (2003).
6. T. B. Jones, *Electromechanics of Particles*, Cambridge: Cambridge University Press, 1995.
7. P. R. C. Gascoyne and J. Vykoukal,. *Electrophoresis* **23**, 1973-1983 (2002).
8. R. Krupke, F. Hennrich, H. v. Löhneysen and M. M. Kappes, *Science* **301**, 344-437 (2003).
9. H. Peng, N. T. Alvarez, C. Kittrell, R. H. Hauge and H. K. Schmidt, *J. Am. Chem. Soc.* **128**, 8396-8397 (2006).
10. R. Gambari, M. Borgatti, L. Altomare, N. Manaresi, G. Medoro, A. Romani, M. Tartagni and R. Guerrieri, *J.Technol. Cancer Res T.* **2**, 31-39 (2003).
11. R. Hölzel, *J. Electrostat.* **56**, 435-447 (2002).
12. R. Hölzel, N. Gajovic-Eichelmann and F. F. Bier, *Biosens. Bioelec.* **18**, 555-564 (2003).
13. N. G. Green, A. Ramos, A. Gonzalez, H. Morgan and A. Castellanos, *Phys. Rev. E* **66**, 026305 (2002).
14. M. Washizu, *J. Electrostat.* **25**, 109-123 (1990).
15. F. Dewarrat, M. Calame and C. Schönenberger, *Single Mol.* **3**, 189-193 (2002).
16. N. G. Green, H. Morgan and J. J. Milner, *J. Biochem. Biophys. Methods* **35**, 89-102 (1997).
17. X.-B. Wang, Y. Huang, R. Hölzel, J. P. H. Burt and R. Pethig, *J. Phys D: Appl. Phys.* **26**, 312-322 (1993).
18. M. Washizu, *J. Electrostat.* **62**, 15-33 (2004).
19. N. G. Green, A. Ramos and H. Morgan, *J. Phys. D: Appl. Phys.* **33**, 632-641 (2000).
20. V. Namasivayam, R. G. Larson, D. T. Burke and M. A. Burns, *Anal. Chem.* **74**, 3378-3385 (2002).
21. W. A. Germishuizen, C. Wälti, R. Wirtz, M. B. Johnston, M. Pepper, A. G. Davies and A. P. J. Middelberg, *Nanotechnology* **14**, 896-902 (2003).
22. A. Ramos, H. Morgan, N. G. Green and A. Castellanos, *J. Electrostat.* **47**, 71-81 (1999).

Parallel Assisted Assembly of Multilayer DNA and Protein Nanoparticle Structures Using a CMOS Electronic Array

Michael J. Heller[1*], Dietrich A. Dehlinger[2], Benjamin D. Sullivan[1]

[1]University of California San Diego, Department of Bioengineering and Department of Electrical and Computer Engineering, La Jolla, CA 92093-0412 USA

[2]University of California San Diego, Department of Electrical and Computer Engineering, La Jolla, CA 92093 USA

*Correspondence to mheller@bioeng.ucsd.edu

Abstract. A CMOS electronic microarray device was used to carry out the rapid parallel assembly of functionalized nanoparticles into multilayer structures. Electronic microarrays produce reconfigurable DC electric fields that allow DNA, proteins as well as charged molecules to be rapidly transported from the bulk solution and addressed to specifically activated sites on the array surface. Such a device was used to carry out the assisted self-assembly DNA, biotin and streptavidin derivatized fluorescent nanoparticles into multilayer structures. Nanoparticle addressing could be carried out in about 15 seconds, and forty depositions of nanoparticles were completed in less than one hour. The final multilayered 3D nanostructures were verified by scanning electron microscopy.

Keywords: DNA, proteins, nanoparticles, self-assembly, electric field, CMOS, microarrays

Introduction

One of the grand challenges in nanotechnology is the development of fabrication technologies that will lead to cost effective nanomanufacturing processes. In addition to the more classical top-down processes such as photolithography, so-called bottom-up processes are also being developed for carrying out self-assembly of nanostructures into higher order structures, materials and devices [1-6]. To this end, considerable efforts have been carried out on both passive and active types of Layer-by-Layer (LBL) self-assembly processes as a way to make three dimensional layered structures which can have macroscopic x-y dimensions [7-24]. In cases where patterned structures are desired, the substrate material is generally pre-patterned using masking and a photolithographic process [25, 26]. Other approaches to patterning include the use of optically patterned ITO films and active deposition of the nanoparticles [27, 28]. Nevertheless, limitations of passive LBL and as well as active assembly processes provide considerable incentive to

CP 859, *DNA-Based Nanoscale Integration: International Symposium,* edited by W. Fritzsche
© 2006 American Institute of Physics 978-0-7354-0357-4/06/$23.00

continue the development of better paradigms for nanofabrication and heterogeneous integration.

Over the past decade electronic microarray devices, produced by a top-down photolithography process, have been developed for DNA diagnostic applications. In these applications, electronic microarray devices which produce reconfigurable electric fields on their surfaces are first used to address and bind negatively charged biotinylated DNA molecules to selected test-sites on the microarray surface. In the next step, samples containing unknown DNA sequences are then rapidly transported and selectively hybridized to the DNA sequences bound at the specific test-sites [29-37]. Thus, these devices are able to direct and accelerate the self-assembly or "bottom-up" process of DNA hybridization occurring on the microarray. In addition to the directed transport and addressing of biomolecules, the ability of electronic microarrays to carry out the rapid patterned deposition of charged nanoparticles was also demonstrated early in the development of the technology [30]. Ultimately, electronic microarrays have been used to carry out transport, addressing and selective binding of a variety of charged biomolecules such as DNA, RNA, biotin/streptavidin, and antibodies [40]; nanoparticles [30, 44]; cells [41] and even 20 micron sized light emitting diode (LED) semiconductor devices [42, 43]. Overall, these devices have proven the ability to manipulate a surprising large variety of entities that range in size from small molecules to micron scale objects [45, 46, 47]. Using a 400 site CMOS microarray device and controller system we have now demonstrated rapid and highly parallel assisted self-assembly of DNA, biotin and streptavidin derivatized nanoparticles into multilayer structures.

Materials and Methods

400 Site CMOS Array Device Specifications - The 400 site CMOS microarray device (Nanogen) is able to independently output currents to each of the 400 different electrodes (54 micron diameter, in 25 columns by 16 rows) at up to 1 μA at 5 volts. Surrounding the 400 test-site microelectrodes are four large counter electrodes which encompass the inner electrode array (Fig 1). The 400 site CMOS array is coated with a 10 μm thick polyacrylamide gel permeation layer which is impregnated with streptavidin. The CMOS microchip array is flip-chip bonded onto a ceramic platform, which then is mounted onto the CMOS controller system. The CMOS controller system is itself mounted under an epifluorescent microscope system with associated CCD camera and imaging system. This allows the nanoparticle addressing and deposition process to be monitored both electrically and optically in real time. The CMOS controller is run by a laptop computer.

Parallel Nanoparticle Addressing (Layering) Process - In order to use the 400 site CMOS array device as a nanomanufacturing platform, standard procedures were developed to determine the optimal parameters for parallel 3D nanoparticle layering. The CMOS array device was initially prepared by first washing it several times with ultra-purified water to remove a protective carbohydrate layer from the permeation layer. After washing, the permeation layer surface was reacted with a 20 μl of a 2 μM Biotin Dextran (Sigma B-9264) solution for 30 minutes. The array was then finally washed with a 100 mM L-histidine solution. To test various addressing (deposition) conditions, the microarray device was programmed to be activated in columns with currents varying from 0.025 µA to 0.4 µA in 0.025 µA increments. Since the nanoparticles used in the experiments carried a net negative charge, the electrodes at the desired addressing sites on the array were biased positive, and the larger counter electrodes on the perimeter of the device were biased negative. Typically a group of alternating columns on the array would be activated in parallel at the different current levels and addressing times, while the intervening columns of electrodes were not activated and thus served as negative controls for the nanoparticle addressing and binding process. Array addressing was carried out using 10 µm of 100 mM L-histidine buffer, containing from 1-10 nM of the derivatized nanoparticles. The two types of derivatized nanoparticles used in the nanoparticle layering experiments being described were 40 nanometer red fluorescent polystyrene nanoparticles derivatized with streptavidin (Molecular Probes F8770, Ex 580 nm - Em 605 nm) and 40 nanometer yellow-green fluorescent polystyrene nanoparticles derivatized with biotin (Molecular Probes F8766, Ex 505 nm – Em 515 nm). The biotin-streptavidin ligand binding reaction allows the two different types of fluorescent nanoparticles to be bound to each other, but nanoparticles of the same type do not specifically bind to each other.

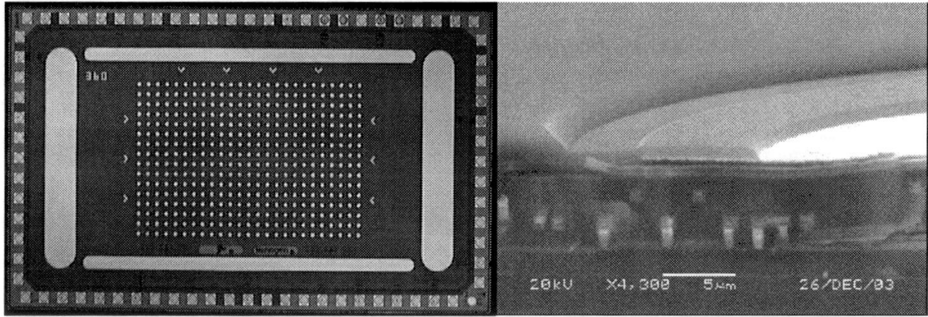

FIGURE 1 – The left side shows photo of a CMOS microarray (5mm x 7mm) with four hundred 50 micron test-site microelectrodes and four large perimeter counter electrodes used to produce electric field geometries that encompass the whole microarray surface area (~ 4mm x 6 mm). The chip is coated with a polyacrylamide gel permeation layer which is one micron thick when dry and about 10 microns thick when hydrated. The permeation layer is impregnated with streptavidin, which then allows biotinylated DNA sequences, proteins, or nanoparticles to be immobilized onto the permeation layer surface. The right side photo shows a cross section through the chip at one of the 50 micron sites exposing the underlying silicon base, some of the CMOS circuitry and a portion of the platinum microelectrode structure.

The nanoparticle addressing, binding and layering experiments were carried out as follows: (1) First Nanoparticle Layer - About 20 µl of a 10 nM solution of 40 nm streptavidin nanoparticles (red fluorescence) in 100 mM L-histidine was placed on the microarray and the selected columns of electrodes were activated at the different current levels (0.025µA -0.4 µA), with addressing times of 5, 15 and 30 seconds. (2) Wash - The array was then immediately washed (manually) three times with 100 mM L-histidine, which usually took less than one minute. Epifluorescence (Ex 580 nm - Em 605 nm) monitoring of the array was carried out during the process. (3) Second Nanoparticle Layer - About 20 µl of a 10 nM solution of 40 nm biotin nanoparticles (green fluorescence) in 100 mM L-histidine was placed on the array and selected electrodes were activated (currents form 0.025-0.4 µA, with addressing times of 5, 15 and 30 seconds). Epifluorescence (Ex 505 nm – Em 515 nm) monitoring of the array was carried out during the process. (4) Wash - The array was then immediately washed three times with 100 mM L-histidine. (5) Successive Nanoparticle Layers - Steps 1-4 were repeated to achieve desired number of nanoparticle layers, and epifluorescence (red and green) monitoring of the array was carried during the whole process. (6) Final Wash - The array was finally washed several times with deionized water to remove L-histidine. Using the process described above, a total of forty alternating addressing and depositions of the 40 nm streptavidin nanoparticles (red fluorescence) and 40 nm biotin nanoparticles (green fluorescence)were carried out in about one hour.

The same process was carried out with DNA derivatized nanoparticles. Biotinylated complementary DNA strands were separately hybridized with streptavidin microspheres in a 50:1 ratio in a solution of 100mM L-histidine and 20 mm NaCl. The sequences used were 5'-Biotin-GAA-CAG-CTT-TGA-GGT-GCG-TGT-TTG-TGC-CTG-TCC-TGG-GAG-AGA-CCG-GCG-CAC-3' (Main sequence) and 5'-Biotin-GTG-CGC-CGG-TCT-CTC-CCA-GGA-CAG-GCA-CAA-ACA-CGC-ACC-TCA-AAG-CTG-TTC-3' (complementary sequence). The chip surface was prepared with streptavidin as described previously. Next the main sequence strand in histidine and NaCl was addressed down to the chip for 0, 5, 15, 30 seconds in 2 regions across the chip for a total of 6 columns for each time, 3 in each region. This created regions of varying DNA activation across the chip. Afterwards, the beads with the complementary DNA sequence were pulled down for 5 and 15 seconds in the 2 respective regions in 2 out of the three columns for each condition. The chip was washed and then the beads with the main sequence DNA were pulled down in the same manner 2 out of the three columns for each condition such that one column was overlapped with the complementary DNA beads for each deposition condition. This was repeated a total of 20 layers.

Characterization Methodology/SEM Analysis of Nanostructures - To directly view the resulting layers, the array device with the layered nanostructures (assembled on the activated 54 micron sites) was sputtered with a 10 nm layer of gold to create a thin conductive layer. The array was then imaged by scanning electron microscopy (SEM). The 40 nm nanoparticles were easily resolvable by SEM and the nature of the deposited material could be observed.

Results

In order to determine optimal conditions for derivatized nanoparticle layering, experiments were carried out at addressing times of 5 seconds, 15 seconds and for 30 seconds. For each of the addressing time experiments, ten columns (16 sites) were activated with DC current levels that ranged from 0.025 µA to 0.4 µA, in increments of 0.025 µA. The activation of all 160 sites (at different current levels) was carried out in parallel. For each addressing time experiment (5 seconds, 15 seconds and 30 seconds) the addressing process was carried out forty times with alternating 40 nanometer red fluorescent streptavidin nanoparticles and green fluorescent biotin nanoparticles. In these experiments, the alternate columns were not activated. Figure 2a shows the fluorescent microscope imaging results for 39 addressings at 5 seconds, Fig 2b shows 39 addressing at 15 seconds and Fig 2c shows the 39 addressings at 30 seconds. The inserts in each of the three photos (2a, 2b and 2c) are enlargements showing four of the 54 µm sites, which were in the rows that received 0.375 µA levels of DC current. The images show no detectable fluorescence is observed on any of the alternate sites which were not activated. By the relative fluorescent intensity of the activated sites, the best conditions for nanoparticle layering appear to be at the 5 second and 15 second addressing times in the 0.30 to 0.40 µA current level range. At the lower current levels (<0.30 µA) the overall fluorescent intensity for the layers begins to decrease. At the longer 30 second addressing time, the nanoparticle layers become visibly damaged. Under real time epifluorescent microscope observation some of these fractured layers could actually be observed to flap when the sites were activated.

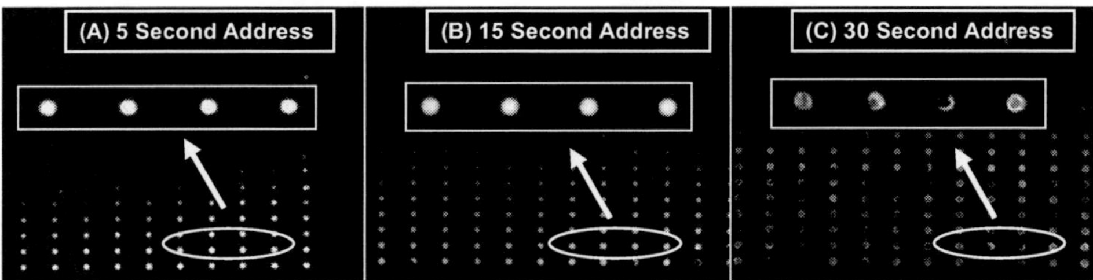

FIGURE 2 - Shows fluorescent analysis results for highly parallel electric field 3D layering of 39 alternate layers of 40 nanometer red fluorescent streptavidin nanoparticles and green fluorescent biotin nanoparticles onto the 50 micron diameter test sites of a CMOS 400 site microarray device. Addressing times were 5 seconds (A), 15 seconds (B), and 30 seconds (C) at DC current levels that ranged from 0.025 uA to 0.40 uA. Inserts are enlargements of a small section of the microarray each showing the results for 0.375 uA on four test sites. Nanoparticle addressing times between 5 to 15 seconds with DC current between 0.35 to 0.40 uA appear to be optimal. The longer 30 second addressing time caused damage to the layered nanostructure. Total layering time is one minute or less per layer, including the washing steps.

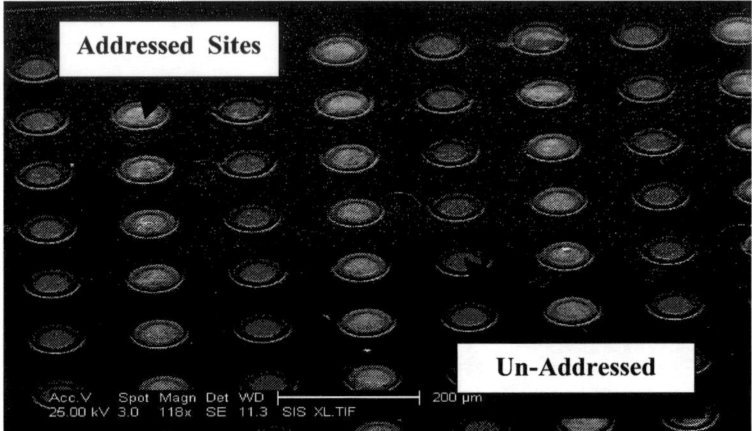

FIGURE 3 - Shows an SEM image of the microarray surface with the alternating rows of the addressed or activated sites (lighter areas) and un-addressed or un-activated sites (darker areas). The activated (light areas) were addressed with 40 nanoparticle depositions at the various current levels. The non-activated sites (darker areas) serve as control for any non-specific nanoparticle accumulation. (Scale in the image is 200 µm).

Scanning electronic microscopy was used to examine the forty layer nanoparticle structures in more detail. Figure 3 shows an SEM image of the microarray surface where the alternating rows of the addressed or activated sites appear as the lighter areas, and un-addressed or un-activated sites (negative control sites) appear as somewhat darker areas. The un-addressed site (negative control site) shows little or no nanoparticle accumulation on the permeation gel layer surface, even though it was exposed forty times to the solutions containing nanoparticles. The SEM image of the addressed site shows a very well defined top layer of DNA nanoparticles that appear to about 40 nanometers in diameter.

Discussion and Conclusions

We have shown that electric field assisted self-assembly of forty layer biotin and streptavidin 40 nm nanoparticle structures, and twenty layer DNA derivatized 40 nm nanoparticles could be carried out in a rapid and highly parallel format using a CMOS electronic microarray device. In this process, efficient nanoparticle addressing/deposition was achieved in 15 seconds or less. With a washing step of about 45 seconds, the total time for creating a nanoparticle layer was about one minute. Thus, the forty layer biotin/streptavidin nanoparticle structures, and the twenty layer DNA derivatized nanoparticle structures could be completed in less than one hour. The optimal electronic addressing window for creating high quality 3D layered structures appears to be at current levels in the 0.25 µA to 0.35 µA range [Fig 4]. Because the nanoparticles used in our experiments were of the same size (40 nm), it is difficult to determine from the SEM image the homogeneity of individual biotin and streptavidin or DNA nanoparticle layers within the multilayered structures. Nevertheless, when taken together with fluorescent monitoring of depositions and the experiments showing no layering of like nanoparticles, the SEM images are consistent with a forty and twenty layer alternating nanoparticle structure. Work is now in progress with different sized nanoparticles to better determine the true quality of individual layers within the multilayer structures.

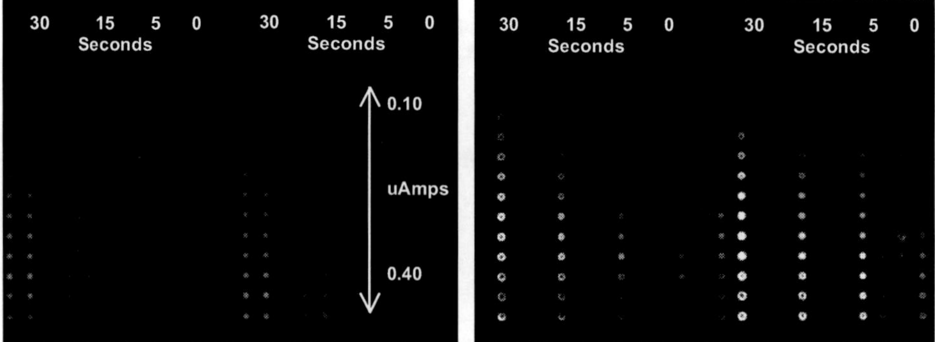

FIGURE 4 - Shows fluorescent analysis results for highly parallel electric field 3D layering of 20 alternate layers of 40 nanometer DNA derivatized red fluorescent streptavidin nanoparticles and green fluorescent nanoparticles onto the 50 micron diameter test sites of a CMOS 400 site microarray device. Addressing times for the DNA derivatized beads were 5 and 15 seconds on the left and right sides of the chip respectively.

The results of these experiments have demonstrated a number of advantages relevant to 3D self-assembly nanofabrication. First, it was shown that high electric field strengths and electrophoretic transport from the bulk solution could be maintained even through many layers of nanoparticles, allowing the successive upper layers of nanoparticles to be continually addressed to the activated sites. Second, our results showed that the integrity of the biotin-streptavidin ligand binding was maintained within the forty layer structures, and with the hybridized DNA nanoparticle structures.. This is extremely important for any 3D self-assembly process that is based on biomolecular binding using highly specific protein or DNA ligands. This ability to use high electric field strength with minimal side effects is likely due to the permeation layer which separates the electrode from the nanostructures, as well as to the relatively short addressing times (5-15 seconds) needed for nanoparticle addressing and deposition. Third, because the high electric field strengths produce enormous concentration effects at the activated sites, only minimal concentrations of nanoparticles were needed for addressing and deposition. In fact, the concentration of nanoparticles used was so low that no significant non-specific binding of nanoparticles occurred during the whole process. Fourth, CMOS electronic array devices appear to allow nanoparticle layering to be carried out much faster (one to two orders of magnitude) than comparable passive Layer-By-Layer (LBL) self-assembly nanofabrication processes. Finally, with regard to x-y patterning during the fabrication process, we believe that CMOS array devices may also have advantages over lithographic and ITO patterning processes. Lithographic patterning of the substrate and subsequent layers would be a relatively time consuming process [25, 26], while electronic microarray patterning could be carried almost instantaneously. In the case of ITO (optical) patterning, the patterning step could be rapid, but ITO substrates are not stable at the high DC electric fields needed for rapid addressing and deposition [27, 28]. Overall, the use of a microelectronic array device for assisted self-assembly represents a unique example of combining "top-down" and "bottom-up" technologies into a potentially useful nanofabrication process. Such a process may be useful for the hierarchal assembly of integrated nano/micro/macrostructures for a variety of electronic, photonic, materials and other applications.

References

1. Prime KL and Whitesides GM. "Self-Assembled Organic Monolayers: Model Systems for Studying Adsorption of Proteins at Surfaces." Science 252 (1991): 1164-7
2. Kim B, Tripp SL, and Wei A. "Self-organization of large gold nanoparticle arrays." J Am Chem Soc 123 (2001): 7955-6.
3. Bowden N, Terfort A, Carbeck J, and Whitesides GM. "Self-assembly of mesoscale objects into ordered two-dimensional arrays." Science 276 (1997): 233-5.
4. Fink J, Kiely CJ, Bethell D, and Schiffrin DJ. "Self-organization of nanosized gold particles." Chem Mater 10 (1998): 922-6.
5. Lee S-W, Mao C, Flynn CE, and Belcher AM. "Ordering of quantum dots using genetically engineered viruses." Science 296 (2002): 892-5.
6. Mirkin CA, Letsinger RL, Mucic RC, Storhoff JJ. "A DNA-based method for rationally assembling nanoparticles into macroscopic materials." Nature 382 (1996): 607-9.

7. "Multilayer Thin Films-Sequential Assembly of Nanocomposite Materials", Decher, D., Schlenoff, J. B., Editors, Wiley-VCH Verlag, Weinheim, Germany, 2003

8. "Nanoelectromechanics in Engineering and Biology", Hughes, M.P, Editor, CRC Press, Boca Raton, FL, 2003.

9. "Handbook of Nanoscience, Engineering and Technology". Goddard, Brenner, Lyashevski & Lafrate, Editors, CRC Press, Boca Raton, FL, 2003.

10. Artyukhin AB, Bakajin O, Stroeve P, Noy A, Layer-by-layer electrostatic self-assembly of polyelectrolyte nanoshells on individual carbon nanotube templates, Langmuir 20 (4): 1442-1448 (2004)

11. Yuan JJ, Zhou SX, You B, Wu LM, Organic pigment particles coated with colloidal nano-silica particles via layer-by-layer assembly, Chemistry of Materials 17 (14): 3587-3594 (2005)

12. Ma N, Zhang HY, Song B, Wang ZQ, Zhang X, Polymer micelles as building blocks for layer-by-layer assembly: An approach for incorporation and controlled release of water-insoluble dyes, Chemistry Of Materials 17 (20): 5065-5069 (2005)

13. Zapotoczny S, Golonka M, Nowakowska M, Novel photoactive polymeric multilayer films formed via electrostatic self-assembly, Macromolecular Rapid Communications 26 (13): 1049-1054 (2005)

14. Wanunu M, Popovitz-Biro R, Cohen H, Vaskevich A, Rubinstein I, Coordination-based gold nanoparticle layers, Journal Of The American Chemical Society 127 (25): 9207-9215 (2005)

15. Hammond PT, Form and function in multilayer assembly: New applications at the nanoscale, Advanced Materials 16 (15): 1271-1293 (2004)

16. Jacobs HO, Campbell SA, Steward MG, Approaching nanoxerography: The use of electrostatic forces to position nanoparticles with 100 nm scale resolution, Advanced Materials 14 (21): 1553 (2002)

17. Barry CR, Gu J, Jacobs HO, Charging process and coulomb-force-directed printing of nanoparticles with sub-100-nm lateral resolution, Nano Letters 5 (10): 2078-2084 (2005)

18. Mardilovich P, Kornilovitch P, Electrochemical fabrication of nanodimensional multilayer films, Nano Letters 5 (10): 1899-1904 (2005)

19. Allred DB, Sarikaya M, Baneyx F, Schwartz DT, Electrochemical nanofabrication using crystalline protein masks, Nano Letters 5 (4): 609-613 (2005)

20. Tsai DH, Kim SH, Corrigan TD, Phaneuf RJ, Zachariah MR, Electrostatic-directed deposition of nanoparticles on a field generating substrate, Nanotechnology 16 (9): 1856-1862 (2005)

21. Lumsdon SO, Kaler EW, Velev OD, Two-dimensional crystallization of microspheres by a coplanar AC electric field, Langmuir 20 (6): 2108-2116 (2004)

22. Lin JC, Yates MZ, Petkoska AT, Jacobs S, Electric-field-driven assembly of oriented molecular-sieve films, Advanced Materials 16 (21): 1944 (2004)

23. Wang N, Lin H, Li J, Yang X, Chi B, Electrophoretic deposition and optical properties of titania nanotube films, Thin Solid Films 496 649-652 (2006)

24. Ferrari B, González S, Moreno R, Baudín C, Multilayer coatings with improved reliability produced by aqueous electrophoretic deposition, J. European Ceramic Society 26 (2006) 27-36

25. Hua F, Shi J, Lvov Y, Cui T, Patterning of layer-by-layer self-assembled multiple types of nanoparticle thin films by lithographic technique, Nano Letters 2 (11): 1219-1222 (2002)

26. Ko HH, Jiang CY, Tsukruk VV, Encapsulating nanoparticle arrays into layer-by-layer multilayers by capillary transfer lithography, Chemistry Of Materials 17 (22): 5489-5497 (2005)

27. Gao M, Sun J, Dulkeith E, Gaponik N, Lemmer U, Feldmann J, Lateral patterning of CdTe Nanocrystal Films by the Electric Field Directed Layer-by-Layer Assembly Method, Langmuir 2002, 18 4098-4102

28. Hayward RC, Saville DA, Aksay IA, Electrophoretic assembly of colloidal crystals with optically tunable micropatterns, Nature Vol 404 56-59 (2000)

29. Heller MJ, Tu E, Martinsons R, Anderson RR, Gurtner C, Forster A, Sosnowski R, 2002, "Active microelectronic array systems for DNA hybridization, genotyping, pharmacogenomics and Nanofabrication Applications", in Integrated Microfabricated Devices, Marcel Dekker, Eds. Heller and Guttman, Chap.10, pp.223-270.

30. Heller M. J., 1996. An Active Microelectronics Device for Multiplex DNA Analysis. IEEE Engineering in Medicine and Biology, 15: 100-103.

31. Sosnowski, R.G., Tu, E., Butler, W.F., O'Connell, J.P., and Heller, MJ, 1997. "Rapid Determination of Single Base Mismatch in DNA Hybrids by Direct Electric Field Control", Proc. Nat Acad. Sci. USA, Vol. 94, pp.1119-23.

32. Edman, C. F., Raymond, D. E., Wu, D. J., Tu, E., Sosnowski, R.G., Butler, W.F., Nerenberg, M., and Heller, MJ, 1997. "Electric Field Directed Nucleic Acid Hybridization On Microchips" Nucleic Acids Research, Vol. 25, #24, pp. 4907-4914.

33. Heller, MJ, "An Integrated Microelectronic Hybridization System for Genomic Research and Diagnostic Applications", in Micro Total Analysis Systems "98, Edited by D. J. Harrison and A. van den Berg, Kluwer Academic Publishers, pp. 221-224, 1998.

34. Heller, MJ, Tu, E., Holmsen, A., Sosnowski, R. G., O'Connell, J. P., (1999) "Active Microelectronic Arrays for DNA Hybridization Analysis" in DNA Microarrays: A Practical Approach, Edited by M. Schena, Oxford University Press, pp. 167-185, 1999.

35. Heller MJ. Forster AH, Tu E, "Active Microelectronic Chip Devices Which Utilize Controlled Electrophoretic Fields for Multiplex DNA Hybridization and Genomic Applications, Electrophoresis 2000, Vol. 21, pp. 157-64

36. Gurtner C, Tu E, Jamshidi N, Haigis R, Onofrey T, Edman CF, Sosnowski R, Wallace B and Heller MJ, "Microelectronic Array Devices and Techniques for Electric Field Enhanced DNA Hybridization In Low-Conductance Buffers", Electrophoresis 2002, 23, 1543-1550.

37. Kassengne SK, Reese H, Hodko D, Yang JM, Sarkar K, Swanson P. Raymond DE, Heller MJ, and Madou MJ, "Numerical Modeling of Transport and Accumulation of DNA on Electronically Active Biochips', Sensors and Actuators 2003, B 94, 81-98.

38. Esener, S.C., D. Hartmann, M. J. Heller and J. M. Cable, "DNA Assisted Micro-Assembly: A Heterogeneous Integration Technology For Optoelectronics," *Proc. SPIE Critical Reviews of Optical Science and Technology, Heterogeneous Integration*, Ed. A. Hussain, CR70, Chapter 7, January 1998.

39. Gurtner C, Edman CF, Formosa RE, Heller MJ. 2000. Photoelectrophoretic Transport and Hybridization of DNA on Unpatterned Silicon Substrates. J. Am. Chem. Soc. 122(36):8589-94

40. Huang Y, Ewalt KL, Tirado M, Haigis R. Forster A, Ackley D, Heller MJ, O'Connell, JP, Krihak M. 2001. Electric manipulation of bioparticles and macromolecules on microfabricated electrodes. Analytical Chemistry (73):1549-59.

41. Cheng, J., Sheldon, E.L., Wu, L., Uribe, A., Gerrue, L.O., Carrino, J., Heller, M. O'Connell J. 1998. Electric field controlled preparation and hybridization analysis of DNA/RNA from E. coli on microfabricated bioelectronic chips. Nature Biotechnology, 16, 541-546

42. Edman CF, Gurtner, C, Formosa RE, Coleman JJ, Heller MJ. 2000. Electric-Field-Directed Pick-and-Place Assembly. HDI. (3)10: 30-35

43. Edman CF, Swint RB, Gurthner C, Formosa RE, Roh SD, Lee KE, Swanson PD, Ackley DE, Colman JJ. Heller MJ, 2000. Electric Field Directed Assembly of an InGaAs LED onto Silicon Circuitry. IEEE Photonics Tech. Letters, 12(9):1198-1200

44. Daniel M. Hartmann, David Schwartz, Gene Tu, Mike Heller, Sadik C. Esener "Selective DNA attachment of particles to substrates " Journal of Materials Research, Vol. 17, No. 2, February 2002, pp. 473-478.

45. US # 6,569,382 "Methods and Apparatus for the Electronic Homogeneous Assembly and Fabrication of Devices", issued May 27, 2003

46. US #6,652,808 "Methods for the Electronic Assembly and Fabrication of Devices", issued Nov. 25, 2003.

47. US #6,706,473 "Systems and Devices for the Photoelectrophoretic Transport and Hybridization of Oligonucleotides", issued March 16, 2004

Acknowledgements

We thank NSF for funding related to NSF award no.DMI-0327077, and Nanogen for funding and supplying CMOS microarrays and the array controller system.

Analysis of G-wire DNA Conductivity

James Vesenka*, Robert Baron†,
Scott D. Collins†, & Rosemary L. Smith†

*University of New England, 11 Hills Beach Rd. Biddeford ME, 04005, USA
†Laboratory for Surface Science and Technology, University of Maine, Orono, ME 04469, USA

Abstract. The electrical conductivity of G-wire DNA which is adsorbed to the surface of mica was examined with the assistance of silicon shadow masks. Four point probe masks were fabricated in silicon using photolithographic patterning and dry reactive ion etching. The silicon "stencils" were designed specifically for use in generating shadow deposited, metal contacts on top of G-wire DNA samples adsorbed to the atomically flat surface of mica. Two types of metal contacts were used in these experiments; electron beam evaporated gold using a high vacuum system and argon sputtered gold using a low vacuum scanning electron microscopy sample coating apparatus. The metal electrode patterning was characterized through atomic force microscopy imaging. The conductivity of the G-wire DNA samples was analyzed using a high impedance multimeter. The lower limit of resistance of the G-wire DNA networks was determined to be in excess of 1 GΩ, indicating from these experiments that G-wire DNA appears to be an insulator.

Keywords: G-wire DNA, G-wire Networks, electrical conductivity, shadow mask
PACS: 73.63.-b, 68.37.Ps

BACKGROUND

The electrical conductivity of *duplex-* DNA has been the subject of numerous recent studies. These experiments were focused on examining the use of dried DNA as a molecular wire. Conductivity through short DNA sequences (~15 base pairs) has been established by photo-induced and flash-quench techniques [1]. Experiments on single stranded DNA have suggested that short segments of DNA (~30 bp, about 10nm) are electrically conducting [2,3,4,5]. However, the majority of experiments on longer strands (>100nm) indicate that duplex DNA is an insulator under dried conditions and low relative humidity [6,7,8,9]. Experiments that have measured conductivity in bundles of DNA suggest that the conduction mechanism may be related to charge migration through hydrated samples [9]. In this paper we examine the conductivity of dried four-stranded "G-wire" DNA.

G-DNA is a polymorphic family of four-stranded structures containing guanine tetrad motifs [10,11]. Guanine rich oligonucleotides that are self-complimentary, as found in many telomeric (chromosome ends) repeat sequences, form G-DNA in the presence of monovalent and/or divalent metal cations. The hairpin structures, comprised of guanine quartets, are thought to play a role in telomerase activity and are? essential for DNA replication [12]. Marsh and Henderson [13] established that

CP 859, *DNA-Based Nanoscale Integration: International Symposium*, edited by W. Fritzsche
© 2006 American Institute of Physics 978-0-7354-0357-4/06/$23.00

self-assembled Guanine-rich tetraplex DNA could *self-assemble* to micrometer lengths, called "G-wires," in large quantities through the overlap of the repeated $G_4T_2G_4$ (Tet1.5) oligonucleotide sequence found in the G-quartets (Fig. 1a) [13]. The growth of these structures is thermodynamically driven [14]. The ionic conditions under which G-wires were grown determined the type of caged metal cations (e.g. Mg^{++}, K^+, or Na^+), integrated into the structure. In contrast to double-stranded DNA, the integrated metal cations surrounded by the four-stranded phosphate backbone, plus extensive hydrogen bonding of the G-quartets, might facilitate lateral conductivity over their uniform 2.4 nm diameter and micrometer-lengths. The uninterrupted nanometer-scale morphology of G-wire DNA would make them an ideal candidate for use in nanotechnology.

Hydration low-current scanning tunneling microscopy (HLCSTM) measurements indicated that G-wires could be reproducibly imaged at tunneling currents above a picoampere at moderate relative humidity (Fig. 1b) [15]. The contrast mechanism for non-conductive molecules imaged by HLCSTM stems in part from the hydration layer on top of a hygroscopic substrate, such as mica in humid air, depending only upon the applied "bias" voltage [16]. Under low bias, high-resolution imaging is maintained by conduction through the hydration layer. At high voltages ballistic tunneling takes place through the air gap into the hydration layer, at the expense of resolution. The HLCSTM observations were interesting because of the factor of 10 to 100 *increase* in tunneling current attained in the presence of G-wire DNA, compared to double stranded DNA. The implication was that the G-wires were assisting conductivity over the substrate. The caged metal cations integrated into a hydrated G-wire molecule may have some mobility. In addition, the metal cations are sufficiently close to each other (estimated between 0.3-0.7 nm) to support electron tunneling. Lastly, the π-bonding of adjacent Guanine-quartets may overlap enough to enhance electron "hopping". All of these mechanisms could also contribute to the lateral conductivity that would make the G-wires a target system for electronic biomolecular devices.

In this paper, the lateral conductivity of G-wires is examined using four point probe measurements. Contacts to the G-wires are made using shadow masked gold films. Instead of adsorbing G-wires onto previously patterned gold electrodes, the gold contacts are made on top of already adsorbed G-wires to ensure good contact between the biomolecules and the metal contacts.

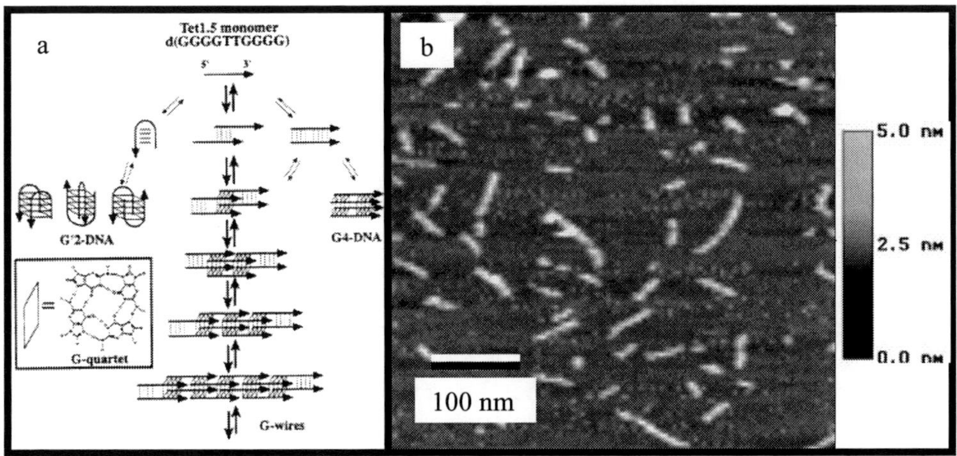

FIGURE 1. (a) Schematic representation of the self-assembly of G-quartet DNA (insert) into multimer G-wire DNA. Monomeric cation species such as potassium or sodium are thought to help stabilize the G-wires between adjacent G-quartets. The thymine groups may act as flexible links that can compress or stretch in solution or after adsorption onto a substrate. (b) Hydration low current scanning tunneling microscopy image of G-wire DNA on the surface of mica taken at 70% relative humidity, 7.5V bias and 3.5 pA tunneling current.

MATERIALS AND METHODS

Quadruplex G-wire DNA was prepared according to the procedure outlined by Marsh et al. [13]. Melting of the $G_4T_2G_4$ monomers (Tet1.5) was maintained with a PCR Thermocylcer (Thermo Hybaid, U.K.), i.e. the growth cocktail and Tet1.5 were is raised to 95°C for ten minutes to promote the melting of fortuitous G-4 structures, i.e. to ensure a monomeric concentration of G-wires. Samples of concentrated G-wires (monomer concentration 1.0 mM) were diluted and incubated for five minutes either directly on freshly cleaved muscovite mica, or on Parafilm® for ten minutes followed by direct adsorption onto mica at room temperature in a buffer consisting of 10 mM Tris (pH 7.6) and 1 mM $MgCl_2$. The samples were then rinsed with 1 ml de-ionized water to remove excess buffer salts. Freshly prepared G-wire DNA was imaged with a Molecular Imaging PicoSPM low current STM under controlled humidity with a Digital Instruments Nanoscope E controller. G-wire DNA were imaged at RH ≈ 60-85%, bias voltages of -5 to -10 V, and tunneling current of ≈ 1-3.5 picoamperes. Silicon shadow masks, or stencils, were made by lithographic patterning and dry reactive-ion etching. The stencil pattern creates four metal contacts, separated by 600 microns, with which a four-point probe measurement can be made to determine the sheet resistance of the film. Metal contacts were made by thermal electron beam evaporation of gold onto the DNA samples at high vacuum (Temescal, 10^{-9} Torr) or by argon sputter coating at low vacuum (Polaron, 10^{-1} Torr). A Nanoscope IIIa controller and Multimode AFM operated in Tapping Mode were used to image the G-wire DNA networks and the integrated metalized contacts. Conductivity measurements were made using both two and four point -probes and a HP 3258A Multimeter, with a maximum resistance measurement of 1.2 GΩ at a potential of 500V and maximum current capacity of 500 nA through the sample.

RESULTS AND DISCUSSION

The metalized contacts were made by evaporating or sputtering gold through shadow masks, such as those shown in Fig. 2a and b. The contacts were deposited (150 nm – Fig. 3a) over a variety of conductive and non-conductive surfaces, including G-wire DNA networks (Fig. 3b). The samples stayed very nearly at room temperature during the coating procedure to ensure no degradation due to heating. The impact of extreme dehydration of the samples under vacuum conditions was not examined. Table 1 represents average conductivity measurements from three samples each of three different types of conductive thin films, control surfaces (mica and mica rinsed and dried with imaging buffer) and six different samples of G-wire DNA networks. The thin film thicknesses were measured with the AFM in tapping mode. No visible surface deposits were imaged on either the mica or buffer treated mica surfaces. G-wire DNA networks were found to be in excellent contact with the gold coatings (Fig. 3b). The calculated values of the resistance are in the same range as the measured resistances found with the multimeter. No observed conductivity was found on the mica and buffer treated mica samples, as expected. Also no measurable conductivity was found over G-wire DNA network samples in either dry or hydrated states. Rehydration of the G-wires was undertaken by exposing the samples to moist air over a saturated salt solution of potassium chloride (85% relative humidity at 20°C) over an hour. The resistance was measured with the probes already connected and inside the enclosed humidity chamber. The time used in this experiment may very well be insufficient to rehydrate the G-wires sufficiently to return the biomolecules to their native solution state. The results appear to be inconsistent with the HLCSTM data, though the reasons for the difference are probably due structure differences between dried and hydrated DNA. In summary the lower limit of G-wire resistance is in excess of 1 GΩ (10^5 Ω-m) under dry conditions.

Sample Conductivities

Sample	Depth (nm)	ρ (Ω-m)	$R_{calculated}$ (Ω)	$R_{measured}$ (Ω)
Gold	13±1	2.5×10^{-9}	1.8	18±5
Chromium	13±1	1.3×10^{-8}	10.	40±5
Carbon	12±1	10×10^{-5}	8300	8700±1000
Freshly cleaved Mica	-	10^9	$>1.2 \times 10^9$	$>1.2 \times 10^9$
Mica rinsed with buffer, water and dried	-	-	$>1.2 \times 10^9$	$>1.2 \times 10^9$
G-wire DNA in either wet or dry air	2.0±0.1	$>10^5$	?	$>1.2 \times 10^9$

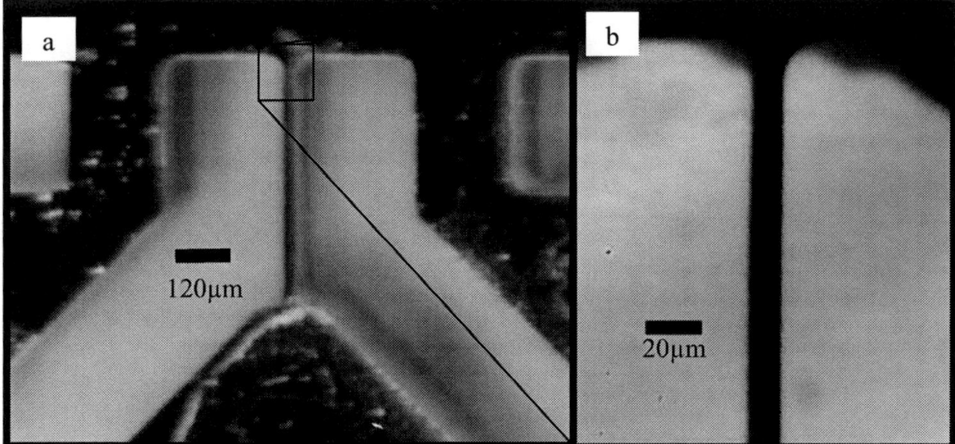

FIGURE 2. (a) Low resolution bright field optical image of four-point probe configuration. The two arms on the wings are for setting up large currents at high voltages in conductive samples. (b) High resolution optical image of gold contacts made by evaporating gold through the shadow mask at left. The gap between the two center probes is about 15 μm and can be used for traditional two point probe experiments. N.B.: the shadow on the top right of the gold contacts is a result of thinner gold coating due to incomplete removal of the silicon oxide layer during the dry etching process.

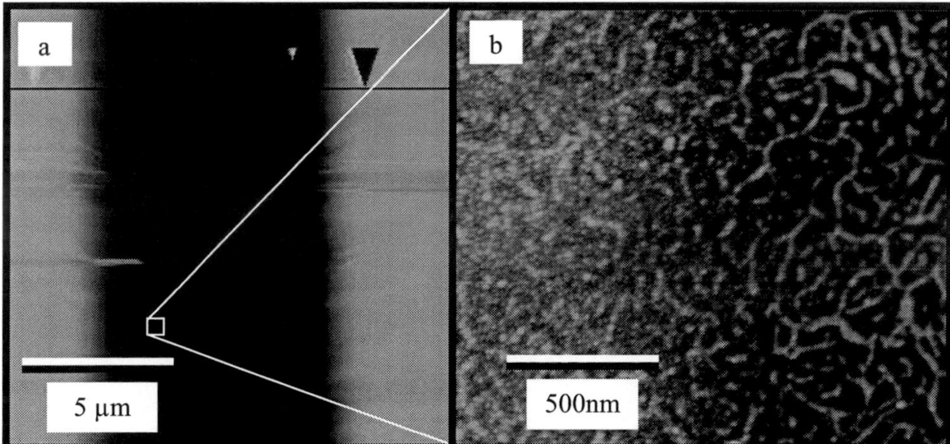

FIGURE 3. (a) Gold contacts on either side of this image are 150 nm above the G-wire DNA networks observed in the background in (b). The transition region between gold contacts and the G-wire DNA networks is shown in (b), where the height scale is 10 nm from dark to bright.

ACKNOWLEDGMENTS

This work was supported by the David and Lucille Packard Foundation, Interdisciplinary Research. The authors thank of Matthias Urban, Robert Kretschmer, and Patrick Spinney for their assistance with initial conductivity measurements.

REFERENCES

1. T.J. Meade & J.F. Kayyem, *Angew. Chem. Int. Engl.* **34**, 352-354 (1995).
2. H.-W. Fink and C. Schönenberger, *Nature* **398**, 407-410 (1999)
3. D. Porath, A. Bezryadin, S. de Vries, C. Dekker, *Nature* **43**, 635-638 (2000).
4. K.-H. Yoo, D.H. Ha, J.-O. Lee, J.W. Park, Jinhee Kim, J.J. Kim, H.-Y. Lee, T. Kawai, and H. Y Choi, *Phys. Rev. Lett.* **87**, 198102-1-198102-4 (2001).
5. H. Cohen, C. Nogues, R. Naaman, and D. Porath, *Proc. Nat. Acad. Sci. U.S.A.* **102**, 11589-11593 (2005).
6. Y.X Zhou, A.T. Johnson Jr., J. Hone, and W.F. Smith, *Nano Lett.,* **3**, 1371. (2000)
7. P.J. de Pablo, M.T. Martínez, J. Colchero, J. Gómez-Herrero, W.K. Maser, A.M. Benito, E. Muñoz, and A.M. Baró, *Adv. Mater.,* **12**, 573-576 (2000).
8. P.J. de Pablo, F. Moreno-Herrero, J. Colchero, J. Gómez-Herrero, P. Herrero, A.M. Baró, P. Ordejón, J.M. Soler, and E. Artacho, *Phys. Rev. Lett.,* **85**, 4992-4995 (2000).
9. Storm, A.J.; van Noort, J.; de Vries, S.; Dekker, C. *Appl. Phys. Lett.,* **79**, 3881 (2001).
10. J.R. Williamson, M.K. Raghuraman, & T.R. Cech, "Monovalent cation-induced structure of telomeric DNA: the G-quartet model." *Cell.* **59**, 871-880. (1989)
11. J.R. Williamson. *Proc. Natl. Acad. Sci. U.S.A.* **90**, 3124-3124 (1993).
12. D. Sen & W. Gilbert, *Biochem.* **31**, 65 (1992).
13. T. Marsh, J. Vesenka, & E. Henderson, *Nucleic Acids Res.,* **23**, 696-700 (1995).
14. J. Vesenka, E. Henderson, & T. Marsh, Ed. W. Fritzsche, *AIP Conference Proceedings* **640**, 109-122 (2002).
15. T. Armstrong, J. Root, and J. Vesenka, Ed. W. Fritzsche, AIP Conference Proceedings 725. pp. 59-64 (2004).
16. Heim, M., R. Steigerwald, & R. Guckenberger, *J. Structural Bio.,* **119**, 212-221 (1997).

Electronic Properties of DNA-Templated Single-Walled Carbon Nanotubes

Huijun Xin, Héctor A. Becerril and Adam T. Woolley

Department of Chemistry & Biochemistry, Brigham Young University, Provo, UT 84602, USA

Abstract. We have localized SWNTs on DNA templates across electrodes and measured the electrical properties of DNA-templated SWNT assemblies. When a DNA-templated SWNT was deposited on top of and bridging electrodes, the measured conductance was comparable to literature values. In contrast, SWNTs with end-on contacts to the sides of electrodes had conductances hundreds of times lower than literature values, probably due to gaps between the SWNT ends and the electrodes. This work provides a novel approach for localizing SWNTs across contacts in a controlled manner, and our results may be useful in the fabrication of nanoelectronic devices with SWNTs as active components or electrical interconnects.

Keywords: DNA-templated nanofabrication, carbon nanotubes, molecular electronics, nanoelectronics, nanowires, self-assembly.
PACS: 73.63.Fg; 73.63.–b; 85.35.–p

INTRODUCTION

The investigation and application of the electrical properties of carbon nanotubes (CNTs) have attracted extensive attention. Various electronic characteristics of different types of CNTs have been explored. Dekker et al. studied the electrical transport properties of individual metallic single-walled carbon nanotubes (SWNTs) [1]. From current versus voltage (I-V) measurements, they found that current increased in a linear manner with voltage, indicative of quantum wire behavior for metallic SWNTs. Lieber et al. demonstrated a correlation between structural defects and conductance in multi-walled carbon nanotubes (MWNTs) [2]. Their results showed that the conductance values of perfect, defect-free nanotubes were an order of magnitude higher than in CNTs with defects. Haddon et al. measured the conductivity of films of as-prepared and purified SWNTs to be 250-400 S/cm [3]. In contrast, the conductivities of octadecylamine- and poly(m-aminobenzenesulfonic acid)-functionalized SWNT films were 2-3 orders of magnitude lower. McEuen et al. measured electrical transport in ropes of SWNTs [4]. They found that the conductance was suppressed near zero voltage at temperatures lower than 10 K, and the gate voltage modulated the number of conduction electrons in the SWNT ropes. These effects were indicative of single-electron charging and resonant tunneling through the quantized energy levels of ropes of SWNTs.

Because of their unique electronic properties, CNTs have been used to fabricate field-effect transistors (FETs) [5-7] and chemical sensors [8]. Avouris et al. constructed FETs using SWNTs and MWNTs [6]. In the SWNT-FET devices, the conductance could be modulated over five orders of magnitude with the gate voltage. The MWNT-FETs showed no gate effects, but structurally deformed MWNTs had FET behavior. Li et al. fabricated a SWNT sensor platform [8], which could sense NO_2 gas and organic vapors (acetone, benzene and nitrotoluene) by detecting changes in SWNT conductance.

The electronic properties of CNTs can be probed using scanning tunneling microscopy [9] or conductive atomic force microscopy (AFM) [10, 11] by scanning a tip along the length of CNTs. The connection of CNTs with microelectrodes is not critical in these scanning probe microscopy measurement methods. Two techniques, which require contacts of CNTs to electrodes, are used more widely for the study of electrical properties of CNTs and their application in devices. The first approach involves depositing a CNT suspension [12] or growing CNTs by chemical vapor deposition (CVD) [3] on substrates, and then fabricating contacts using lithography techniques. This method generates high-quality contacts between CNTs and electrodes but requires sophisticated techniques (e.g., electron beam lithography) to pattern the contacts. The second approach proceeds in the opposite order: a dispersed SWNT suspension is deposited [13] or SWNTs are grown [14] on substrates on which contacts have already been fabricated. In this strategy, the electrodes are easier to make, but the positions of the CNTs are more difficult to control, and contact quality varies significantly.

Previously, we have localized SWNTs on DNA templates on Si using two approaches. The first utilized a bifunctional compound (1-pyrenemethylamine) bridging between DNA and SWNTs [15]. In the second method, a cationic surfactant, dodecyltrimethylammonium bromide (DTAB), was used to disperse SWNTs in aqueous solution and localize them on DNA templates through electrostatic interactions [16]. In this work, we localized DTAB-wrapped SWNTs on DNAs, which were prealigned across microelectrodes, and the I-V properties of individual DNA-templated SWNTs were measured. In this technique, the placement of SWNTs depends on the positions of the prealigned DNAs. Importantly, this procedure provides a simple way to localize SWNTs across electrodes in a controlled manner, which may be useful in the fabrication of nanoelectronic devices using SWNTs.

EXPERIMENTAL

The gold microelectrodes were fabricated using standard UV lithography techniques. An electrode layout was made using computer-aided design software and then patterned on a glass-chromium photolithographic mask using monochromatic UV light in an Electromask (TRE, Santa Ana, CA) pattern generator. Next, a 500-nm-thick thermal oxide layer was grown on p-type silicon <100> wafers (TTI Silicon, Sunnyvale, CA) using a tube furnace (Bruce Technologies, Billerica, MA). The wafers were coated with a 300-nm-thick layer of AZ-3312 positive photoresist (AZ Electronic Materials, Somerville, NJ) using a spinner (Laurell, North Wales, PA) and soft baked for 60 s at 90 °C prior to exposure. The photoresist-coated wafers were exposed to monochromatic UV light using the step-and-repeat mode of the pattern generator for improved resolution of small features in the electrode pattern. Exposed wafers were soft baked at 110 °C for 60 s and immersed for 60 s in AZ-300 developer solution (AZ

Electronic Materials). Exposed silicon dioxide regions on the wafers were cleaned with a ~10-s dip in buffered oxide etch (Baker, Phillipsburg, NJ), and the substrates were placed in a CHA-600 triple-source thermal evaporator (CHA, Fremont, CA). Approximately 5 nm of Cr and 35 nm of Au were deposited on the wafers, and the excess metal and photoresist were removed from the patterns by an acetone lift-off step, followed by a 30-s dip in freshly prepared piranha solution (a 7:3 mixture of concentrated sulfuric acid and 30% hydrogen peroxide).

SWNT (Carbon Nanotechnologies, Houston, TX) suspensions were prepared in 1% aqueous DTAB solution following a previously developed method [16]. An electrode-patterned substrate was incubated with 40 µL of 10 µg/mL aqueous poly-L-lysine solution for 5 min. A 1-µL droplet of double-stranded λ DNA solution (10 µg/mL) was translated across the electrodes to align DNA [17]. The surface was then treated with fresh SWNT suspension for 10 min. Between each step, the substrate was rinsed thoroughly with deionized water and dried under a stream of nitrogen.

All substrates were studied by AFM. Images were taken with a Multimode IIIa AFM setup (Veeco, Sunnyvale, CA) using aluminum-coated silicon tips (Budgetsensors, Sofia, Bulgaria). Vibrational noise was damped with an active isolation system (MOD1-M, Halcyonics, Goettingen, Germany). AFM imaging parameters were as follows: tip resonance frequency, ~300 kHz; free oscillation amplitude, 0.9-1.1 V; set-point, 0.3-0.9 V; scan rate, 0.4-1.6 Hz. All images were processed after acquisition to eliminate background curvature.

The I-V curves were measured at room temperature under ambient conditions using a system combining an E5262 two-channel high-speed source monitor unit (Agilent, Palo Alto, CA) with a low-noise probe station (Micromanipulator Inc., Carson City, NV). The scan range was from −0.1 to +0.1 V for SWNT samples and from −0.5 to +0.5 V for control surfaces. Tungsten probe tips, with a point radius of 0.5 µm, contacted the microelectrodes.

RESULTS AND DISCUSSION

Figure 1 shows a photograph of the gold microelectrodes that DNA (and SWNTs) were aligned across for I-V measurements. The electrodes were designed with gaps ranging from small (<500 nm) to large (~500 µm). From 35 substrates measured, the smallest distance between the A/B electrodes and the C electrode was between 200 and 1,000 nm, with a considerable fraction (~50%) between 450-550 nm, which was in the range of typical suspended SWNT lengths [16]. This setup was designed such that individual SWNTs could be localized bridging the electrodes on prealigned surface DNA in the small-gap area.

DNA molecules were deposited directionally across the space between the A/B and C electrodes. If the electrode surfaces were >10 nm higher than the Si, DNA deposition was sparse. In this study, the electrodes were ~6 nm taller than the Si surface, which provided adequate coverage of DNA on the substrate. **Figure 2A** and **Fig. 2D** show AFM data of DNA aligned on the surface across electrodes.

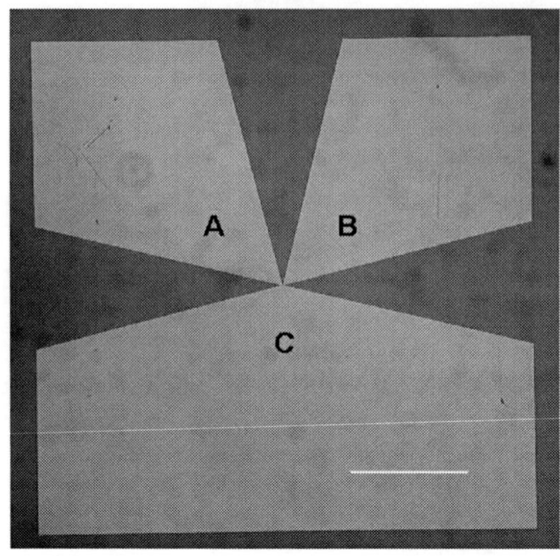

FIGURE 1. Photograph of the microelectrode design. A, B, and C are lithographically patterned gold microelectrodes. The scale bar represents 500 μm.

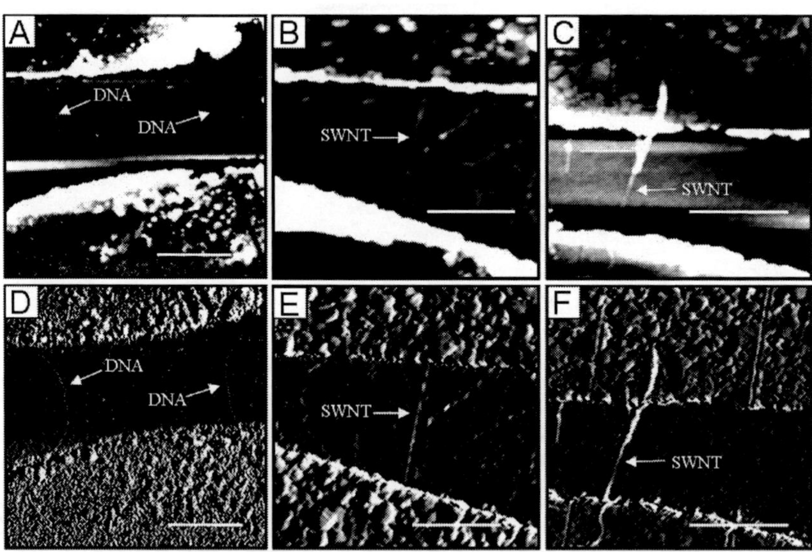

FIGURE 2. AFM height (A-C) and amplitude (D-F) images of DNA and DNA-templated SWNTs deposited on electrodes. (A, D) DNA aligned across electrodes prior to SWNT localization. (B, E) A DNA-templated SWNT between electrodes. (C, F) A DNA-templated SWNT on top of and bridging electrodes. The scale bars represent 500 nm in all images. DNAs and SWNTs are labeled by arrows.

AFM images in **Fig. 2B-C** and **Fig. 2E-F** present individual SWNTs localized on prealigned DNA traversing electrodes. On most substrates, SWNT deposition was similar to that shown in **Fig. 2B** and **Fig. 2E**, where a SWNT was on the surface between electrodes, but did not have obvious contact with the tops of the electrodes. In an atypical SWNT placement, shown in **Fig. 2C** and **Fig. 2F**, a SWNT was deposited on top of both electrodes and bridging the gap.

I-V properties of various surfaces are reported in **Fig. 3**. The I-V curves in **Fig. 3A** were from two control surfaces: an electrode substrate prior to any treatments and a substrate with DNA aligned across electrodes and which was treated with 1% DTAB lacking SWNTs. The conductance of these control surfaces was $<10^{-12}$ S. The I-V curves in **Fig. 3B-C** were from the SWNTs in **Fig. 2B-C**. The conductance of the SWNT in **Fig. 2B** was 1.7×10^{-8} S. The SWNT with contacts to the tops of the electrodes in **Fig. 2C** had a conductance of 2.2×10^{-6} S, which was >200 times higher than the one in **Fig. 2B**. Substrates with SWNT placement similar to that in **Fig. 2B** had conductance values in the range of 7.3×10^{-9}-4.8×10^{-8} S (see **Table 1**). It appears that the contact resistance is considerably lower when SWNTs are located on top of the electrodes, compared to when SWNTs are placed between electrodes.

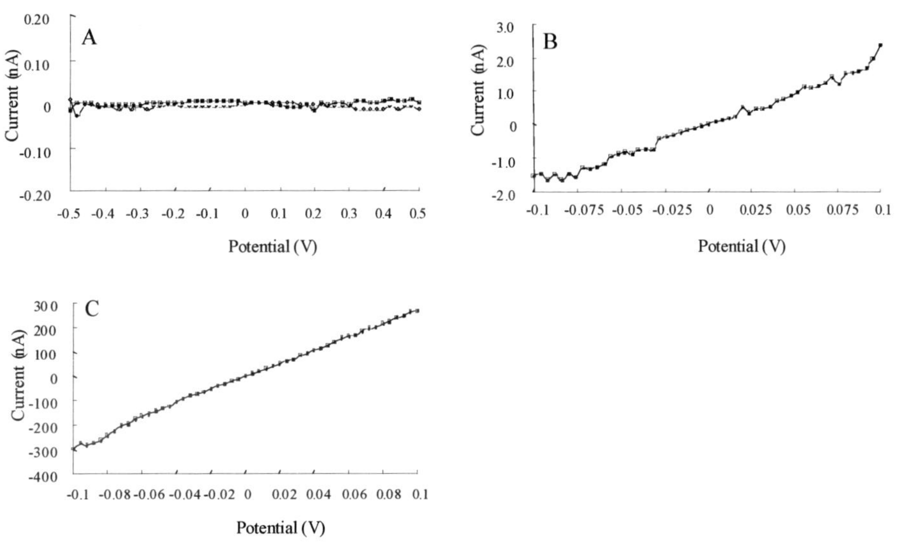

FIGURE 3. I-V curves from control substrates and DNA-templated SWNTs across electrodes. (A) Control substrates: (—◇—) an untreated electrode surface; (—■—) a substrate with aligned DNA that was treated with 1% DTAB lacking SWNTs. (B) A substrate with a SWNT between electrodes (**Fig. 2B**). (C) A substrate with a SWNT on top of and bridging electrodes (**Fig. 2C**).

TABLE 1. Conductance of DNA-templated SWNTs on five substrates. Substrate 5 is the one depicted in **Fig. 2C**.

Substrate	Length (nm)	Resistance (Ω)	Conductance (S)
1	730	1.4×10^{8}	7.3×10^{-9}
2	500	1.0×10^{8}	1.0×10^{-8}
3	560	5.9×10^{7}	1.7×10^{-8}
4	680	2.1×10^{7}	4.8×10^{-8}
5	410	4.5×10^{5}	2.2×10^{-6}

With perfect contacts and two electron transport channels, the conductance of a metallic SWNT should be $4e^2/h$ = 155 μS (e is the electron charge and h is Planck's constant), which corresponds to a resistance of 6.45 kΩ [18]. Imperfect connections to the electrodes add a contact resistance (R_c) to the total resistance [19]. Individual SWNT resistances of ~1 MΩ [20] and 2.9 MΩ [6] at room temperature have been reported. With improved contacts made using titanium, Dai et al. reported a resistance of 11 kΩ for an individual SWNT at room temperature [18]. In this work, when a SWNT was on top of the electrodes, the resistance was 450 kΩ, which is slightly less than the values reported by Tans [20] and Avouris [6], but much higher than the literature value for a SWNT with almost perfect electrical contacts [18]. The other substrates, where SWNTs bridged electrodes, but lacked obvious contacts to the tops of the electrodes, had 100-500 fold higher resistances, which indicates the possibility of small gaps between the SWNT ends and the electrodes.

CONCLUSION

This work demonstrates that DNA-templated SWNTs can be formed across electrodes on surfaces, providing a new approach for localizing SWNTs across contacts in a controlled manner. When a DNA-templated SWNT was placed on top of and bridging electrodes, the measured conductance was comparable to literature values. On the other hand, SWNTs that lacked contacts to the tops of electrodes had conductance values hundreds of times lower than those in the literature, probably due to gaps between SWNTs and electrodes. Our approach for controlled surface placement of SWNTs may be useful in making nanoelectronic or nanomechanical devices.

In the future, it will be critical to optimize the connections between SWNTs and electrodes. One potential improvement would be to have the electrodes level with the Si surface, so there is no height difference between the electrodes and Si. In this way, the DNAs should deposit on top of the electrodes more readily, and SWNTs should thus be localized on top of electrodes, as in **Fig. 2C**. In addition, metals such as Ti have a low CNT contact resistance [18] and should improve conductance properties. An alternative method to solve the contact problem would be to deposit metal into the gaps between SWNTs and electrodes to decrease the contact resistance. These improvements should move the measured electrical properties of SWNTs closer to theoretical values.

ACKNOWLEDGMENTS

Microfabrication work was performed in the Integrated Microelectronics Laboratory at Brigham Young University. This work was also supported in part by the Army Research Laboratory and the U.S. Army Research Office under grant number DAAD19-02-1-0353.

REFERENCES

1. Venema, L. C.; Wildöer, J. W. C.; Janssen, J. W.; Tans, S. J.; Tuinstra, H. L. J.; Kouwenhoven, L. P.; Dekker, C. *Science* **1999**, *283*, 52-55.
2. Dai, H.; Wong, E. W.; Lieber, C. M. *Science* **1996**, *272*, 523-526.
3. Bekyarova, E.; Itkis, M. E.; Cabrera, N.; Zhao, B.; Yu, A.; Gao, J.; Haddon, R. C. *J. Am. Chem. Soc.* **2005**, *127*, 5990-5995.
4. Bockrath, M.; Cobden, D. H.; McEuen, P. L.; Chopra, N. G.; Zettl, A.; Thess, A.; Smalley, R. E. *Science* **1997**, *275*, 1922-1925.
5. Bradley, K.; Cumings, J.; Star, A.; Gabriel, J.-C. P.; Grüner, G. *Nano Lett.* **2003**, *3*, 639-641.
6. Martel, R.; Schmidt, T.; Shea, H. R.; Hertel, T.; Avouris, Ph. *Appl. Phys. Lett.* **1998**, *73*, 2447-2449.
7. Auvray, S.; Derycke, V.; Goffman, M.; Filoramo, A.; Jost, O.; Bourgoin, J.-P. *Nano Lett.* **2005**, *5*, 451-455.
8. Li, J.; Lu, Y.; Ye, Q.; Cinke, M.; Han, J.; Meyyappan, M. *Nano Lett.* **2003**, *3*, 929-933.
9. Collins, P. G.; Zettl, A.; Bando, H.; Thess, A.; Smalley, R. E. *Science* **1997**, *278*, 100-103.
10. Bockrath, M.; Markovic, N.; Shepard, A.; Tinkham, M.; Gurevich, L.; Kouwehoven, L. P.; Wu, M. W.; Sohn, L. L. *Nano Lett.* **2002**, *2*, 187-190.
11. Heo, J.; Bockrath, M. *Nano Lett.* **2005**, *5*, 853-857.
12. Misewich, J. A.; Martel. R.; Avouris, Ph.; Tsang, J. C.; Heinze, S.; Tersoff, J. *Science* **2003**, *300*, 783-786.
13. Derycke, V.; Martel, R.; Appenzeller, J.; Avouris, Ph. *Nano Lett.* **2001**, *1*, 453-456.
14. Biercuk, M. J.; Mason, N.; Marcus, C. M. *Nano Lett.* **2004**, *4*, 1-4.
15. Xin, H.; Woolley, A. T. *J. Am. Chem. Soc.* **2003**, *125*, 8710-8711.
16. Xin, H.; Woolley, A. T. *Nanotechnology* **2005**, *16*, 2238-2241.
17. Woolley, A. T.; Kelly, R. T. *Nano Lett.* **2001**, *1*, 345-348.
18. Kong, J.; Yenilmez, E.; Tombler, T. W.; Kim, W.; Dai, H. *Phys. Rev. Lett.* **2001**, *87*, 106801-106804.
19. McEuen, P. L.; Fuhrer, M.; Park, H. *IEEE Trans. Nanotechnol.* **2002**, *1*, 78-85.
20. Tans, S. J.; Devoret, M. H.; Dai, H.; Thess, A.; Smalley, R. E.; Geerligs, L. J.; Dekker, C. *Nature* **1997**, *386*, 474-477.

MICROSCOPIC

CHARACTERIZATION

Comparative Study of Atomic Force Imaging of DNA on Graphite and Mica Surfaces

Dmitry Klinov[†,‡], Benjamin Dwir[,†], Eli Kapon[†], Natalia Borovok[§], Tatiana Molotsky[§], Alexander Kotlyar[*,§,**]*

[†]*Laboratory for the Physics of Nanostructures, Ecole Polytechnique Fédérale de Lausanne, CH-1015 Lausanne, Switzerland*

[‡]*Shemyakin-Ovchinnikov Institute of Bioorganic Chemistry, Russian Academy of Sciences, 117871, Moscow, Russia*

[§]*Department of Biochemistry, George S. Wise Faculty of Life Sciences, Tel Aviv University, Ramat Aviv, 69978 Israel*

[**]*Nanotechnology Center, Tel Aviv University, Ramat Aviv, 69978 Israel*

Abstract. Various DNA-based structures (single-, double-, triple-stranded and quadruplex-DNA) were characterized using non-contact atomic-force microscopy on two substrates: modified highly-oriented pyrolitic graphite (HOPG) and mica. Deposition on mica, a conventional substrate used in studies of bio-molecules, results in strong deformation of all above types of molecules while deposition on modified HOPG affects the morphology of DNA much less compared to mica. This is demonstrated by a larger measured height of the DNA molecules deposited on HOPG, as compared to mica, and an increased flexibility of the molecules, evidenced by a shorter molecular end-to-end distance on HOPG. The estimated heights of the triplex and the quadruplex DNA measured on HOPG are similar to the diameter of these molecules in liquid. We thus conclude that modified HOPG is a substrate more suitable than mica for AFM characterization of DNA morphology.

Keywords: Atomic force microscopy, DNA imaging, poly(dG)-poly(dC), poly(dG)-poly(dG)-poly(dC), triplex DNA, quadruplex DNA, HOPG surface modification.
PACS: 87.14.Gg, 87.15.-v, 87.15.La, 87.64.Dz, 68.37.Ps

INTRODUCTION

Since the first X-ray diffraction (XRD) images of DNA, which formed the basis for the Watson-Crick Model [1], the study of the structure of DNA has been the object of continually improving research techniques. The more recent development of scanning probe techniques, especially scanning tunneling (STM) and atomic force microscopy (AFM) [2], enabled the observation of single molecules, unlike the large number of molecules needed for XRD. However, AFM observations, done usually in air ambient but not limited to it, are always of molecules adsorbed on a surface. Under these circumstances, the forces between the DNA molecule and the surface can not be neglected, and indeed they can largely influence the shape of the molecule as some

CP 859, *DNA-Based Nanoscale Integration: International Symposium,* edited by W. Fritzsche
© 2006 American Institute of Physics 978-0-7354-0357-4/06/$23.00

measurements have already shown [3]. Specifically, images of DNA adsorbed on mica, which is the substrate most commonly used for AFM imaging of DNA, show a molecule height much lower than the diameter of the molecule as measured by NMR [4] or X-ray diffraction [5]. This is attributed to a compression of the molecule caused by its interaction with the substrate [6]. The strong deformation of the molecules caused by deposition on mica thus limits the use of AFM imaging techniques for studies of the molecular structure of DNA. However, there are only few experiments on AFM imaging of DNA on other substrates, like atomically flat graphite [7] and gold surfaces [8], mainly due to the difficulty of DNA deposition on these surfaces.

We developed a novel, low-force AFM measurement technique that overcomes most of the above-mentioned difficulties. It includes the use of a low-force substrate (modified HOPG) and low-amplitude non-contact (NC) mode imaging. We performed comparative AFM characterization of the morphology, the molecular length, height and end-to-end distance, of single-, double-, triple- and four-stranded DNA, on mica and on modified HOPG. The data show that the morphology of DNA is much less affected by interaction with graphite as compared to mica. We believe that the use of our technique opens the way to better characterization of the morphology of DNA through more accurate imaging.

EXPERIMENTAL SECTION

The DNA probes used in the framework of this work were: single-stranded M13mp18 phage DNA of length 7.25 kb (Sigma-Aldrich biotechnology, Switzerland), double-stranded linear pSK+ plasmid DNA, of length 3 kbp. Poly(dG)-poly(dC) duplex, poly(dG)-poly(dG)-poly(dC) triplex and G4 monomolecular quadruplex DNA, composed of 4000 base pairs, 1000 triads and 400 tetrades, respectively, were prepared as described in our recent publications [9,10,11].

The substrates used were: modified HOPG (NT-MDT, Russia) and non-modified freshly cleaved moskovite mica (Distrelec, Switzerland).

The DNA molecules were deposited on HOPG as follows: 40 µl of graphite modifier (GM) solution (Nanotuning, Chernogolovka, Russia) was dropped onto the freshly cleaved HOPG surface and incubated for 10 min at room temperature. This treatment results in the formation of a polymer layer on the graphite surface. The excess of the modifier was removed from the surface by a stream of N_2 gas. 10 µl of 5 nM (in molecules) DNA solution in 2 mM Tris-Acetate buffer (pH 7.0) were applied onto the modified graphite surface and incubated for 10 min at room temperature. The DNA solution was then removed from the surface and the surface was dried under a stream of nitrogen gas.

The DNA molecules were deposited on mica as follows: 10 µl of 1-2 nM solution of DNA in 2 mM Tris-Acetate buffer (pH 7.0) containing 5 mM $MgCl_2$ were dropped onto the freshly cleaved mica surface and incubated for 10 min at room temperature. The surface was then rinsed by distillate water and dried under a stream of N_2 gas.

The samples were imaged by Nanoscope III (Veeco, USA), Ntegra (NT-MDT, Russia) and XE-100 (PSIA, Korea)AFMs. All AFM images were obtained in the intermittent-contact (tapping) mode. The images were "flattened" (each line of the image was fitted to a 2nd-order polynomial, and the polynomial was then subtracted from the image line) by the AFM's image processing software to eliminate image distortions caused by sample tilt and scanner non-linearity. Height, length, and end-to-end distance analyses were performed using a statistical image analysis program (DNACalc, NT-MDT, Russia) and the SPIP analysis software (Image Metrology, Denmark).

RESULTS AND DISCUSSION

Mica is the most commonly used substrate in AFM studies of DNA [3] despite the existence of strong deforming adhesive forces, occurring between the molecules and the surface. As a consequence, the height of double-stranded DNA measured by AFM on mica [3] (0.6-0.7 nm) is much lower than the nominal diameter of the molecule in solution (~2 nm) [4,5]. Indeed, the height of both the pSK+ plasmid DNA (Fig. 2a) and the double-stranded poly(dG)-poly(dC) (Fig. 3a) deposited by us on mica and measured by AFM is approximately equal to 0.7 nm. Recently, successful AFM imaging of double- and single-stranded DNA deposited onto non-modified graphite were demonstrated [12]. However, such deposition results in strong aggregation of the DNA, induced by interaction of the highly hydrophilic nucleic acid with the hydrophobic HOPG surface. In our study, we found neither single nor aggregated DNA molecules on the surface of non-modified graphite. A method of HOPG modification using plasma has been reported recently [13]. The modification resulted in efficient adsorption of DNA and other biopolymers onto the surface. For the present study, we used chemical modification of the graphite surface with a commercially available graphite modifier (GM). This is an aqueous solution of a homopolymer, in which the repeating unit is composed of a hydrophobic hydrocarbon chain and a hydrophilic polypeptide-containing amino group. The hydrocarbon chain of the polymer adheres to the hydrophobic graphite surface and the hydrophilic part carrying the positive charge is exposed to the solution, promoting binding of the negatively charged DNA molecules. The advantage of this chemical modification over plasma modification of HOPG is that it is fast, simple, and does not require special equipment. Moreover, the GM-modified graphite surface is flat, showing a residual roughness of 0.1-0.2 nm in AFM images (can be seen in Figs. 1b and 2b).

The estimated height of both plasmid and poly(dG)-poly(dC) dsDNA molecules deposited on modified graphite (see Figures 2b, 3b) is equal to 1.2 nm. This value is larger than that obtained on mica; however it is smaller than the diameter of the double-stranded DNA in solution.

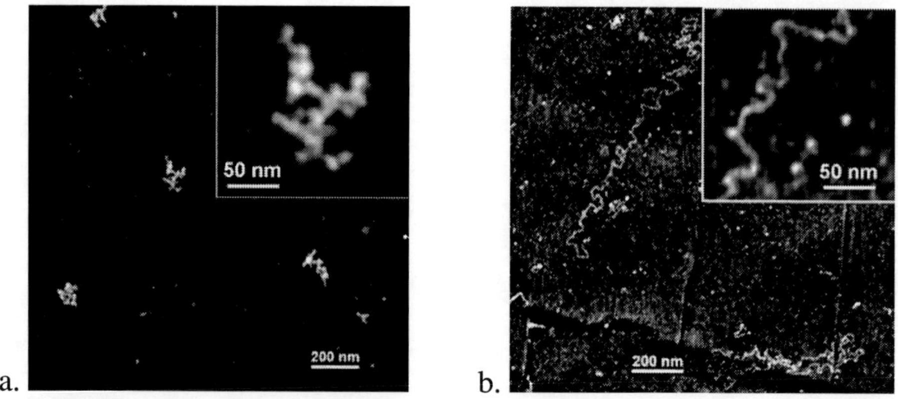

FIGURE 1. AFM images of single-stranded Phage M13 DNA (length 7.25 kb) on mica (a) and on modified HOPG (b). The insets show magnified images.

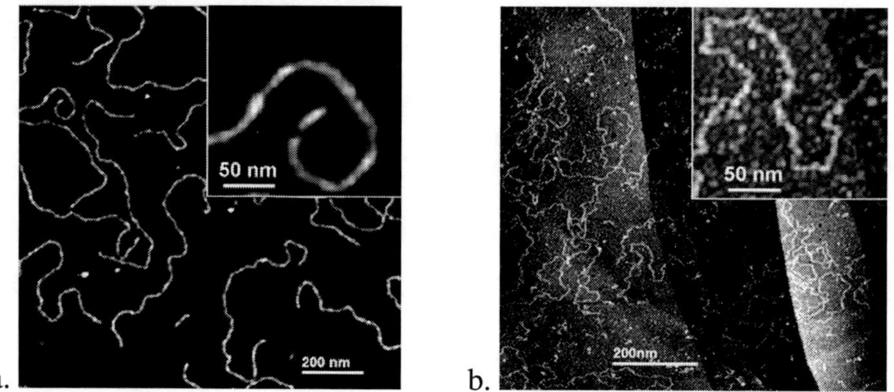

FIGURE 2. AFM images of double-stranded pSK+ plasmid DNA (length 3 kbp) on mica (a) and on modified HOPG (b). The insets show magnified images.

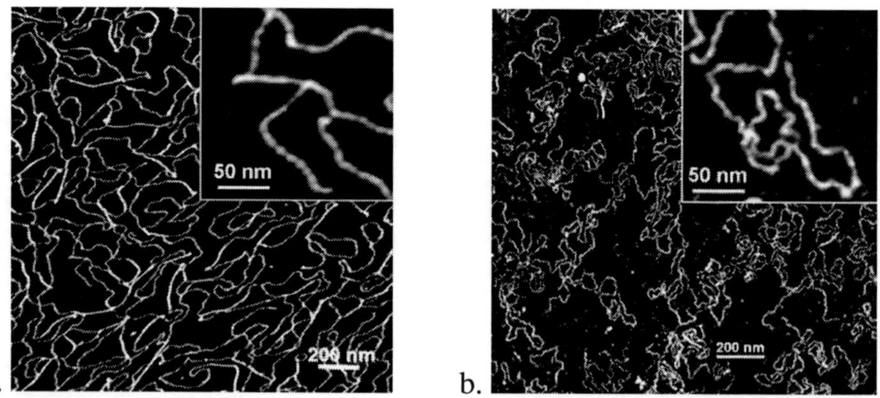

FIGURE 3. AFM images of double-stranded poly(dG)-poly(dC) DNA (length 4 kbp) on mica (a) and on modified HOPG (b). The insets show magnified images.

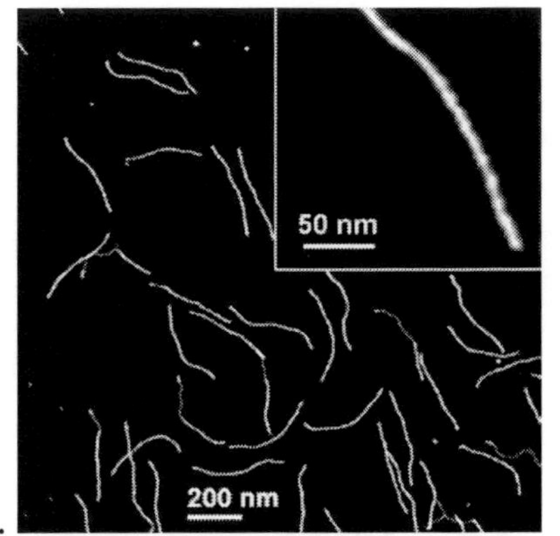 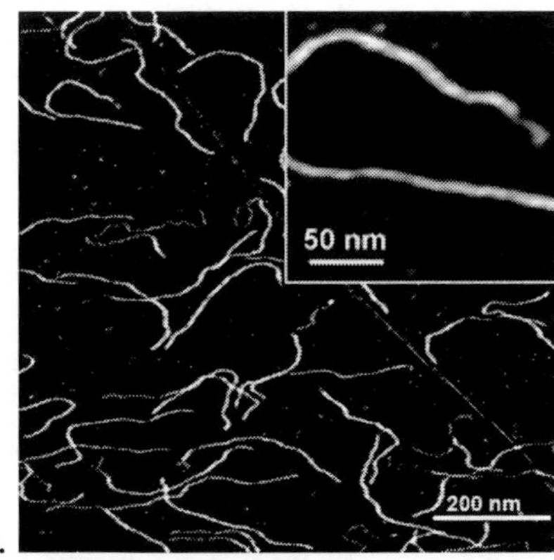

FIGURE 4. AFM images of triple-stranded poly(dG)-poly(dG)-poly(dC) DNA, composed of 1000 triads, on mica (a) and on modified HOPG (b). The insets show magnified images.

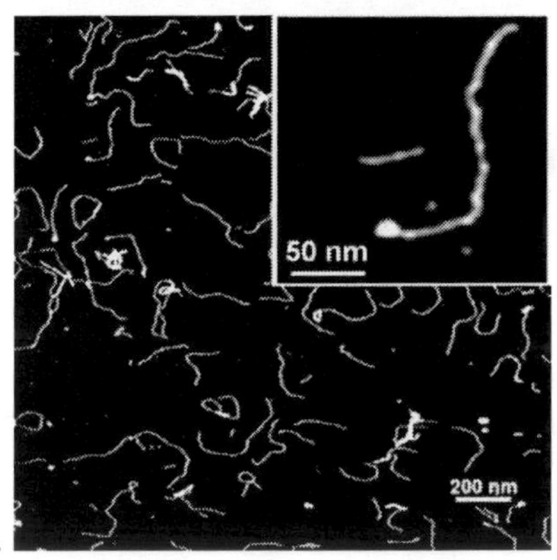 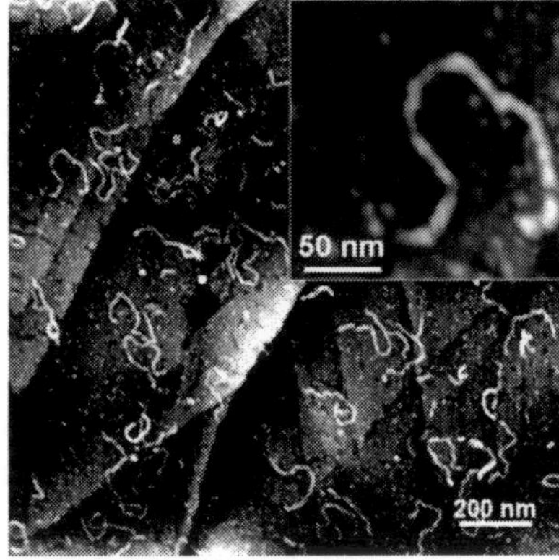

FIGURE 5. AFM images of G4 DNA, composed of 400 tetrads, on mica (a) and on modified HOPG (b). The insets show magnified images.

We have used the same experimental strategy as described above to compare the morphology of DNA-based molecules other than double-stranded ones. To image single-stranded DNA we used the M13 phage DNA (Fig. 1). As can be seen in Fig. 1a, deposition of M13 DNA on mica results in intramolecular aggregation of the DNA; the molecules are seen as compact structures, and their height and length cannot be determined. In contrast to mica, the molecules on HOPG (Fig. 1b) are not aggregated and their height and length can be estimated using AFM. The height of the single-stranded fragments is approximately equal to 0.4 nm, which is about twice as low as that of the double-stranded DNA (~ 1 nm, see Fig. 6).

Figure 2 shows the AFM images of the double-stranded linear plasmid (pSK+) DNA deposited on mica (Fig. 2a) and graphite (Fig. 2b). The apparent morphology of these molecules is very similar to that of the homopolymer, poly(dG)-poly(dC) (Fig. 3). The heights of DNA on mica or on graphite are identical for both molecules, showing that our imaging technique works equally well for natural (random-sequence) as for synthesized DNA.

Figure 4 displays the AFM images of poly(dG)-poly(dG)-poly(dC) triplexes. The height of the molecules measured on modified graphite (Fig. 4b) is 2±0.1 nm, which is equal to the diameter of the triplex structures in solution [14,15]. The increased apparent height of the triplexes with respect to the double-stranded DNA (2nm versus 1 nm, see Fig. 6) reflects the higher stiffness of the former molecules and their resistance to the surface adhesion forces and to the pressure of the AFM tip. The height of the same molecules deposited on mica is approximately 1.2 nm. Our data show that the triplex DNA is more rigid than the double-stranded DNA and resists the interaction with the modified graphite surface and with the AFM tip, thus preserving its native conformation. However, the molecules are not strong enough to resist the forces arising as a result of their interaction with mica.

Figure 5 shows AFM images of G4-DNA. The morphological parameters obtained with G4-DNA are similar to those of triplex DNA (Fig. 3). The height of the G4-DNA molecules deposited on graphite is equal to 2±0.1 nm (Fig. 6). The same molecules that have been deposited on mica show a height of ~1.2 nm (Fig. 6). These data confirm that as in the case of the triplexes, interaction with modified graphite is weaker than with mica and does not affect the molecular morphology of the G4-DNA.

The height of ~ 100 molecules of each type of DNA imaged by AFM (as in Figs. 1-5) was measured, and their average height is plotted in Figure 6 as function of the number of strands composing each molecule. It is clearly seen that the height of all DNA types is larger if the molecules are deposited on HOPG. Moreover, the height of the triplex- and of the G4-DNA is comparable to the diameter of the molecules in solution [14,15,16]. This shows that modified HOPG effects morphology of various DNA types much less than mica.

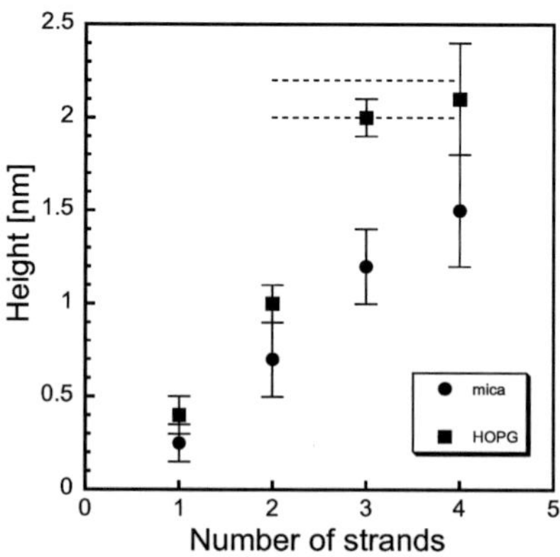

FIGURE 6. Dependence of molecular height on the number of strands composing the DNA molecule. The height of single-stranded M13, double-stranded poly(dG)-poly(dC), triple-stranded poly(dG)-poly(dG)-poly(dC) and four stranded G4-DNA deposited on mica (circles) or on modified HOPG (squares) was measured by AFM. Each data point represents the average height of ~ 100 molecules; the error bars represent the value of 1 SD. The dashed lines show the range of diameter of DNA molecules, as measured in solution [14, 15].

The flexibility of adsorbed DNA molecules is another parameter that has been studied earlier [17], mostly on mica, as a model for polymer gyration [18]. However, the flexibility is governed by intrinsic properties of the DNA molecule, as well as by the surface forces. In order to characterize DNA flexibility, we calculated the stiffness parameter S, defined as the ratio of the measured distance between the opposite ends of the DNA molecule to its entire contour length. This parameter varies from S = 1 for completely stiff, stick-shaped molecules, to $S \approx 0$ for completely folded, compact molecules. We compared the stiffness of poly(dG)-poly(dC), segments of pSK+ plasmid DNA and poly(dG)-poly(dG)-poly(dC), 1000 base pairs and 1000 triads long, respectively, deposited on mica and on HOPG. The results of this analysis, performed on about 100 molecules of each type, are summarized in Table 1. As seen from the table, the triplex molecules are characterized by a larger stiffness compared to the double-stranded DNA, regardless of the substrate used. Both types of double-stranded DNA are characterized by similar stiffness (the somewhat higher observed stiffness of the pSK+ might stem from the use of image sections of longer molecules). Moreover, all types of DNA appear to be stiffer on mica than on graphite. This is probably due to a stronger interaction of the DNA molecules with mica as compared to HOPG. This suggestion is consistent with the results presented in Fig. 6, showing much stronger effect of mica as compared to HOPG on the molecular morphology of DNA.

TABLE 1. Stiffness of double-stranded plasmid DNA, double-stranded poly(dG)-poly(dC), and triplex poly(dG)-poly(dG)-poly(dC) DNA, deposited on mica and on HOPG. Values are an average over 100 molecules (or molecular segments for the plasmid), the uncertainties represent the SD values.

DNA type	Stiffness on mica	Stiffness on HOPG
Plasmid	0.8±0.1	0.4±0.2
Poly(dG)-poly(dC)	0.7±0.2	0.3±0.2
Triplex	0.96±0.03	0.8±0.2

The data presented here clearly demonstrate that modified HOPG has a high affinity to DNA and efficiently binds natural and synthetic DNA molecules. We have shown that the height of all types of molecules tested here is higher, and that their stiffness is lower, if the DNA molecules are deposited on modified graphite rather than on mica. This shows that the molecular morphology of DNA is much less affected by the modified graphite as compared to mica. We thus conclude that modified graphite is a better substrate for AFM characterization of DNA morphology than the more conventional substrate, mica.

ACKNOWLEDGMENTS

This work is supported by a European Grant for Future & Emerging Technologies (IST-2001-38951). We thank NT-MDT for the loan of their AFM.

REFERENCES

1. J. D. Watson, F. H. C. Crick, *Nature* **171**, 737-738 (1953).
2. G. Binning, C. F. Quate, Ch. Gerber, *Phys. Rev. Lett.* **56**, 930-933 (1986).
3. H. G. Hansma, I. Revenko, K. Kim, D. E. Laney, *Nuc. Acids Res.* **24**, 713-720 (1996).
4. N. Tjandra, S. Tate, A. Ono, M. Kainosho, A. Bax, *J. Am. Chem. Soc.* **122**, 6190-6200 (2000); V. R. Parvathy, S. R. Bhaumik, K. V. R. Chary, G. Govil, K. Liu, F. B. Howard, H. T. Miles, *Nuc. Acids Res.* **30**, 1500-1511 (2002).
5. R. E. Franklin, R. Gosling, *Acta Cryst.* **6**, 673-685 (1953); W. Fuller, M. H. F. Wilkins, H. R. Wilson, L. D. Hamilton, *J. Mol. Biol.* **12**, 60-76 (1965).
6. A. Kasumov, D. Klinov, P.-E. Roche, S. Gueron, H. Bouchiat, *Appl. Phys. Lett.* **84**, 1007-1009 (2004).
7. N. Severin, J. Barner, A. Kalachev, J. P. Rabe, *Nano Lett.* **4**, 577-579 (2004).
8. M. Hegner, P. Wagner, G. Semenza, *FEBS Lett.* **336**, 452-456 (1993).
9. A. B. Kotlyar, N. Borovok, T. Molotsky, L. Fadeev, M. Gozin, *Nuc. Acids Res.* **33**, 525-535 (2005).
10. A. Kotlyar, N. Borovok, T. Molotsky, D. Klinov, B. Dwir, E. Kapon, *Nuc. Acids Res.* **33**, 6515-6521 (2005).
11. A. Kotlyar, N. Borovok, T. Molotsky, H. Cohen, E. Shapir, D. Porath, *Adv. Mater.* **17**, 1901-1905 (2005).
12. A. M. O. Brett, A. M. C. Paquim, *Bioelectrochemistry* **66**, 117-124 (2005).
13. D. Klinov, L. P. Martynkina, V. Yu. Yurchenko, V. V. Demin, S. A. Streltsov, Yu. A. Gerasimov, Yu. Yu. Vengerov, *Russ. J. of Bioorg. Chem.* **29**, 363-367 (2003).
14. S. Rhee, Z. Han, K. Liu, H. T. Miles, D. R. Davies, *Biochemistry* **38**, 16810-16815 (1999).
15. D. Vlieghe, L. V. Meervelt, A. Dautant, B. Gallois, G. Précigoux, O. Kennard, *Science* **273**, 1702-1705 (1996).
16. J. T. Davies, *Angew Chemie* **43**, 668-698 (2004).
17. C. Rivetti, M. Guthold, C. Bustamante, *J. Mol. Biol.* **264**, 919-932 (1996).
18. F. Valle, M. Favre, P. De Los Rios, A. Rosa, G. Dietler, *Phys. Rev. Lett.* **95**, 158105(1-4) (2005).

APPENDICES

PROGRAM

Thursday, May 18

20:00 Get-together Buffet
  (Scala Restaurant, 28th floor of Intershop tower,
  5 minutes walk from Hotels, see downtown map)

Friday, May 19

07:45 Pick-up from Hotel IBIS (please meet at front desk)
  public bus to IPHT at Campus Beutenberg

08:30 Opening   (IPHT at Campus Beutenberg)

Biomolecular Approaches (W. Fritzsche)

08:40   **N. Seeman** (New York)
  Using Structural DNA Nanotechnology To Organize Matter
09:05   **A. Bates** (Penn State Univ.)
  RNA-based Nanoscale Integration
09:30   **L. Fruk** (Dortmund)
  *Covalent Hemin-DNA Adducts for Generating a Novel Class
    of Artificial Enzymes*
09:55   **U. Sivan** (Haifa)
  Transistor in a Test Tube and Antibodies to Semiconductors

---- coffee break / posters ----

DNA Manipulation (M. Heller)

10:45   **W. Fritzsche** (Jena)
  DNA-based Approaches for Micro-Nano Integration
11:10   **C. Escude** (Paris)
  *Sequence-specific DNA functionalization for manipulation
    and detection of single DNA molecules*
11:35   **B. Samori** (Bologna)
  DNA adsorption on Inorganic Surfaces and Nanostructure Growth
12:00   **I-M. Hsing** (Hong Kong)
  Integrated Bioanalytical Microdevice Utilizing Nanoparticles

---- lunch break / posters ----

DNA Superstructures (G. Zuccheri)

13:30   **A. Kotlyar** (Tel Aviv)
  Enzymatic Synthesis of Novel Triplex DNA Nanostructures
13:55   **J. Vesenka** (Univ. New England)
  Auto-orientation of G-wire DNA Networks
14:20   **H. Yan** (Tempe)
  *Experimental Progress on DNA-based Self-assembly
    of Nanostructures*

---- coffee break / posters ----

Friday, May 19 (cont.)

DNA–Functionalization (F. Bier)

15:15   **A. Woolley** (B.-Young-Univ.)
*DNA-templated Construction of Metal, Semiconductor
and Biological Nanostructures*
15:40   **A. Pike** (Newcastle) **cancelled!**
*Ferrocenyl-Modified DNA: Synthesis and Characterization of
12-mer Oligonucleotides in Solution and after Integration
with Semiconductor Electrodes*
16:05   **G. Burley** (Munich)
*Directed DNA Metallisation: Towards the Construction of
Rationally Designed, Conductive Nano Devices*
16:30   **A. Filoramo** (Gif/Yvette)
Self-assembly of Carbon Nanotubes by Bio-directed Approach

17:30 Napoleon 1806 Battleground Excursion (bus/hike)

19:30 Dinner ("Alt Jena" Marketplace)

Saturday, May 20

Electrical Manipulation & Measurements (J. Vesenka)

08:30   **M. Heller** (San Diego)
*Parallel Assisted Self-Assembly of Multilayer DNA and Protein
Nanoparticle Structures Using a CMOS Electronic Array*
08:55   **R. Hölzel** (Potsdam)
*Monitoring Dielectrophoretic Collection of DNA by
Electrical Impedance Spectroscopy*
09:20   **H.-Y. Lin** (Taiwan)
*Stretching and Positioning of Individual DNA Molecules
on Electrodes by the Electrohydrodynamic Force*
09:45   **J. Toppari** (Jyväskylä)
*Dielectrophoresis of nanoscale dsDNA using metallic and CNT
electrodes: Humidity effects on dsDNA's electrical conductivity*
10:10   **D. Porath** (Jerusalem)
*Electrical Transport, Polarizability and Spectroscopy
Measurements through DNA Molecules and Derivatives
Using Conductive AFM and STM*

---- coffee break ----

Novel Approaches & Modern Techniques (B. Samori)

10:50   **G. Braun** (Santa Barbara)
Biomolecular Routes for Nanoelectronics
11:15   **D. Klinov** (Lausanne)
High-resolution Atomic-Force Imaging of DNA
11:40   **H. Stadler** (Veeco Instruments)   **cancelled !**
*Manipulation of Biological Surfaces at the Nanoscale:
Prerequisites, Developments, Examples*
12:05   **N. Amro** (NanoInk)
*DNA Nanoarrays Printed via Dip Pen Nanolithography
for High Sensitivity Bioanalysis*

appr. 12:30 End of the scientific program

12:30 Lunch

PHOTOS

Talks
(*Friday, May 19*)

N. Seeman (New York)

A. Bates (Liverpool)

L. Fruk (Dortmund)

U. Sivan (Haifa)

W. Fritzsche (Jena)

C. Escude (Paris)

B. Samori (Bologna)

I.-M. Hsing (Hong Kong)

A. Kotlyar (Tel Aviv)

J. Vesenka (Uni. New England)

M. Heller (San Diego)

G. Braun (Santa Barbara)

A. Woolley (B.-Young-Univ.)

G. Burley (Munich)

A. Filoramo (Gif/Yvette)

Excursion
(*Friday, May 19*)

Talks
(*Saturday, May 20*)

R. Hölzel (Potsdam)

H.-Y. Lin (Taiwan)

J. Toppari (Jyväskylä)

D. Porath (Jerusalem)

H. Yan (Tempe)

D. Klinov (Lausanne)

N. Amro (NanoInk)

PARTICIPANTS

Amro, Nabil	NanoInk	namro@nanoink.net
Bates, Andy	Liverpool	a.d.bates@liverpool.ac.uk
Bergner, Andreas	LOT-ORIEL	bergner@lot-oriel.de
Bier, Frank	Potsdam	frank.bier@ibmt.fraunhofer.de
Björk, Per	Linköping	perbj@ifm.liu.se
Braun, Gary	Santa Barbara	gbraun@chem.ucsb.edu
Brucale, Marco	Bologna	marco.brucale@unibo.it
Burley, Glenn	Munich	Glenn.Burley@cup.uni-muenchen.de
Csaki, Andrea	Jena	andrea.csaki@ipht-jena.de
Escude, Christophe	Paris	escude@mnhn.fr
Festag, Grit	Jena	grit.festag@ipht-jena.de
Filoramo, Arianna	Gif/Yvette	Arianna.Filoramo@cea.fr
Fritzsche, Wolfgang	Jena	wolfgang.fritzsche@ipht-jena.de
Fruk, Ljiljana	Dortmund	ljiljana.fruk@uni-dortmund.de
Gurevich, Leonid	Aalborg	Lg@nanobio.aau.dk
Hamedi, Mahiar	Linköping	mahiar@ifm.liu.se
Hammond, Dave	Munich	dave.hammond@cup.uni-muenchen.de
Hannewald, Karsten	Jena	hannewald@ifto.physik.uni-jena.de
Heller, Mike	San Diego	mheller@bioeng.ucsd.edu
Herland, Anna	Linköping	anher@ifm.liu.se
Hölzel, Ralph	Potsdam	ralph.hoelzel@ibmt.fhg.de
Hsing, I-Ming	Hong Kong	kehsing@ust.hk
Klinov, Dmitry	Lausanne	Dmitry.Klinov@epfl.ch
Kotlyar, Alexander	Tel Aviv	s2shak@post.tau.ac.il
Leiterer, Christian	Jena	c.leiterer@gmx.de
Li, Pei-Yun A.	Munich	jclin6924@yahoo.com.tw
Lin, Chen-an	Munich	jclin6924@yahoo.com.tw
Lin, H.-J.	Taipei	hungyi@mail.nutn.edu.tw
Lyonnais, Sebastien	Gif/Yvette	slyonnai@drecam.cea.fr
Mi, Yongli	Hong Kong	keymix@ust.hk
Mir, Kalim	Oxford	kalim@well.ox.ac.uk
Möller, Robert	Jena	robert.moeller@ipht-jena.de

Müller, Joachim	Dortmund	joschi@Chemie.Uni-Dortmund.DE
Ngyen, Khoa	Gif/Yvette	ttknguyen@cea.fr
Ortmann, Franka	Jena	ortmann@ifto.physik.uni-jena.de
Porath, Danny	Jerusalem	porath@chem.ch.huji.ac.il
Preuß, Martin	Jena	preuss@ifto.physik.uni-jena.de
Rabe, Kersten	Dortmund	kersten.rabe@uni-dortmund.de
Rant, Ulrich	Munich	rant@wsi.tum.de
Reiß, Edda	Potsdam	edda.reiss@ibmt.fraunhofer.de
Ritter, Kathrin	Jena	agathe13@gmx.de
Rossi, Luca	Bologna	Luca.Rossi@bo.infn.it
Samori, Bruno	Bologna	samori@alma.unibo.it
Schüler, Thomas	Jena	thomas.schueler@ipht-jena.de
Seeman, Ned	New York	ncs1@scires.acf.nyu.edu
Sivan, Uri	Haifa	phsivan@tx.technion.ac.il
Steinbrück, Andrea	Jena	andrea.steinbrueck@ipht-jena.de
Streiff, Stephane	Gif/Yvette	stephane.streiff@cea.fr
Toppari, Jussi	Jyväskylä	jussi.toppari@phys.jyu.fi
Vesenka, James	Univ. of New England	jvesenka@une.edu
Wackerbarth, Hainer	Lyngby	hw@kemi.dtu.dk
Wolff, Andreas	Jena	andreas.wolff@ipht-jena.de
Woolley, Adam	Brigham Young Univ.	awoolley@chem.byu.edu
Yan, Hao	Tempe	Hao.Yan@asu.edu
Zuccheri, Giampaolo	Bologna	giampaolo.zuccheri@unibo.it